COLLECTED WORKS OF
RICHARD J. CHORLEY

Volume 2

DIRECTIONS IN GEOGRAPHY

DIRECTIONS IN GEOGRAPHY

Edited by
RICHARD J. CHORLEY

Routledge
Taylor & Francis Group

LONDON AND NEW YORK

First published in 1973 by Methuen and Co. Ltd

This edition first published in 2019
by Routledge
2 Park Square, Milton Park, Abingdon, Oxon OX14 4RN

and by Routledge
52 Vanderbilt Avenue, New York, NY 10017

Routledge is an imprint of the Taylor & Francis Group, an informa business

British Library Cataloguing in Publication Data
A catalogue record for this book is available from the British Library

ISBN: 978-0-367-22096-9 (Set)
ISBN: 978-0-429-27321-6 (Set) (ebk)
ISBN: 978-0-367-22103-4 (Volume 2) (hbk)
ISBN: 978-0-367-22176-8 (Volume 2) (pbk)
ISBN: 978-0-429-27329-2 (Volume 2) (ebk)

Publisher's Note
The publisher has gone to great lengths to ensure the quality of this reprint but points out that some imperfections in the original copies may be apparent.

Disclaimer
The publisher has made every effort to trace copyright holders and would welcome correspondence from those they have been unable to trace.

Directions in Geography

Edited by

RICHARD J. CHORLEY

METHUEN AND CO LTD

First published 1973 by
Methuen & Co Ltd, 11 New Fetter Lane, London EC4
IBM set by Santype Ltd (Coldtype), Salisbury
Printed in Great Britain by
Butler & Tanner Ltd, Frome and London
© 1973 Methuen & Co Ltd

SBN 416 60830 2 (Hardback edition)
416 60840 x (Paperback edition)

Distributed in the USA by
HARPER & ROW PUBLISHERS, INC.
BARNES & NOBLE IMPORT DIVISION

Contents

Preface

In its inception *Directions in Geography* was designed to be more retrospective than has materialized. A decade or so after the flood tide of the 'quantitative revolution' which produced the 'new geography' it was thought valuable to ask a number of geographers, most of whom in one way or another had had some hand in these innovations, to address themselves to their essential nature. It should have been apparent, however, that to expect innovators to engage in introspection was a vain belief, and it is clear from the contributions to this volume that the authors have been dominantly concerned with the many possible directions which our discipline may follow in the future, rather than with the single path which we have trodden up to the present. Clearly, what has happened to geography of recent years gives some pointers to its future, but it is interesting how free our contributors have been in suggesting the fruitful directions which geographical thinking is taking and may well take. In the terminology adopted by one of them, we are not here concerned merely with predicting our future courses of action by 'polishing the present'. Consequently in this volume will be found criticisms of such sacred cows of the 'new geography' as the quantitative approach, human ecology, the academic rationale of geographical work, and of the widespread concern with functioning contemporary regional systems. The fact that geography does not merely possess one certain future but many possible ones is emphasized by the diversity of the views and interests exhibited by the contributors to this book. A counter-point to our fundamentally theoretical needs is provided by calls for utility and relevance, to ecological generalizations by the uniqueness of man's socio-economic systems, to spatial economic efficiency by the requirements of social morality, and to educational practicabilities by an understanding of the true needs of education.

Thus this volume is not in any sense a manifesto, and if it has any single message it is that the recent surge of vitality in our discipline has led to a healthy proliferation of attitudes and objectives which stands in striking contrast to the state of our art before the recent revolution which is so obviously already ancient history to the present authors.

Richard J. Chorley
Sidney Sussex College
Cambridge

PART I

Theoretical

1 · A paradigm for modern geography

BRIAN J.L. BERRY

Time is like a river made up of the events which happen, and its current is strong; no sooner does anything appear than it is swept away, and another comes in its place, and will be swept away too.

Marcus Aurelius Antoninus, Meditations. IV, 43

Several converging threads of dissatisfaction with certain elements in contemporary geography led me to write this essay. In thinking about the implications of spatial field theories I had become increasingly frustrated with what had all too rapidly become "traditional" statistical geography — with the mindless use of conventional inference statistics and measures of association in geographic research without regard for the validity of their assumptions. Many statistical manipulators were ignoring what Dacey was showing clearly enough in the point pattern case: that static pattern analysis is incapable of indicating which of a variety of equally-plausible but fundamentally different causal processes had given rise to the patterns he was studying. This finding seemed generalizable to many other circumstances. At the same time, the "new" environmentalists — today's ecoactivists — were pointing out how irrelevant it is to theorize about the uniform plain so dear to location theorists. Behavioral geographers were calling for, but not producing, new types of theory. New insights were showing the importance of perceptual filters in decision-making. Phenomenology was indicating the limitations of "the" scientific method in the human sciences, a point all the more telling as large-scale planned intervention has appeared in the attempt to guide social change and modify the systems of concern.

What this paper represents is a personal effort to come to terms with the many sources of confusion and doubt about the continuing viability of earlier research orientations that arise from these sources. It owes much to insights derived from John Platt's 1970 paper on "Hierarchical Growth", as well as to the lessons of several years of extra-university involvement in public affairs, confronted by immediate questions of locational and environmental decision-making on the part of city and national governments – i.e. by the real challenges of social relevance. Geography has been characterized as a "mosaic within a mosaic", essentially pluralistic because of the persistence of a variety of strands arising from diverse origins and changing philosophies (Mikesell 1969). It is tempting, therefore, to suggest that any personal statement simply adds another pluralistic element. I have come to feel that there is a unity that transcends the apparent disunity, however, and that the essence of this unity, when distilled, will be the basis for more general geographic theory that will call forth reevaluation and substantial rethinking of the partial theories within each of the separate strands. I view the personal statement as a first step in deriving the more general theory.

The plan of the essay is first to spell out the nature of my dissatisfaction with the state of inference in statistical geography. The conclusions of this critique lead naturally to philosophic preference for "process metageography", and this in turn to a paradigm of locational and environmental decision-making in complex systems that I favor as a guiding orientation for the next generation of geographic research.

Galton's problem: a ghost that haunts statistical geography

The basic problem, from which all other sections of this essay derive was and is the realization that the determinants of spatial variation may be such as systematically to violate one of the most basic assumptions of the conventional inferential procedures that most geographers rely upon. To clarify, implicit in most uses of correlation, regression, factor and like modes of analysis is the assumption that the observations used in the analysis are independent entities for which certain functionally necessary causal relationships between variables occurring within them are equally and generally true. An equally plausible rival hypothesis at the same aggregate level of generalization is, however, that the observations are elements set within larger systems from which they acquire common characteristics by borrowing or migration, or more generally, through the operation of some spatial diffusion mechanism. To say this is to say that much statistical analysis in geography may be confounded by what has now come to be called "Galton's Problem" by American anthropologists (Hildreth and Naroll 1972).

Sir Francis Galton raised his problem at the meeting of the Royal Anthropological Institute in 1889 when Tylor read his pioneering paper introducing the cross-cultural survey method (Tylor 1889, 272). Tylor showed correlations ("adhesions", he called them) between certain traits; in the discussion which followed Galton pointed out that traits often spread by diffusion — by borrowing or migration. Since this is often so, how many independent trials of his correlation did Tylor have?

Galton's problem, then, is to distinguish the effect of functional associations ("adhesions" in Tylor's graphic term) from the effect of more common historical association through diffusion. By a functional association is meant a relationship between one or more variables such that the presence of any one of these tends to facilitate the occurrence of any and all of the others *within* any given area. By diffusion is meant a process involving the acceptance over time of some specific idea or practice (or a set of them, either simultaneously or in sequence) by individuals, groups or other adopting units linked to specific channels of communication, to a social structure, and to a given system of values or culture (Katz, Levin and Hamilton 1963), producing growth that does not appear universally at any one time but manifests itself at points or poles of growth and diffuses in definite channels *between* areas (Perroux 1955).

Franz Boas, for decades the immensely influential dean of American anthropologists, once told his student Lowie (Lowie 1946, 227-30) that when he first read Tylor's paper, he became greatly enthusiastic. The cross-cultural survey method seemed to him an ideal research technique. On reflecting further, however, Galton's objection seemed to him a devastating one; unless there was a solution to Galton's problem, Boas considered the cross-cultural survey method valueless.

The same may be said about much statistical geography. Consider, for example, a study of the votes cast for a 1968 open housing referendum in Flint, Michigan, published in the *Annals of the Association of American Geographers* in 1970 (Brunn and Hoffman 1970). As part of their analysis, the authors used the voting precincts of the city as observational units, and they regressed percent favorable vote separately for the black and white residential areas on median incomes, median housing values, median school years completed and distance from the ghetto core. Ignoring the problem of multicollinearity of income, education, housing value and race, they built into the problem an implicit assumption of functional necessity. First, to be able to predict the vote, their acceptance of a regression framework implied that all one needs to do for any precinct is to measure the variables said to operationalize the causes of voting preferences and to solve the resulting regression equation. Such usage of a regression equation implies that it replicates a uniform causal sequence that arises equally and independently within each of the units of observation. The model thus suggests that voting preference is the result of an evolutionary sequence of events – a unilinear combination of causes – that unfolds homogeneously within each unit of observation. Each precinct, for the purposes of such a model, therefore might well be a similarly-structured but a self-contained spaceship – a closed system – floating in a vacuum totally unrelated to other precincts.

But what if the variables used to construct the model of voting behavior are not related in functionally necessary causal sequences, but merely correlated because they are common outcomes of an institutionally structured dual housing market in which housing choices involve comparison of the relative advantages of a wide range of neighborhoods, subject to externally imposed racial constraints, and if voting in certain precincts is affected by their location relative to other precincts of the opposite kind in the dual system? The causal assumptions of the regression model then fail, as does the assumption that the precincts are independent closed systems, and the particular statistical model is clearly inappropriate.

If the variables are symptoms of phenomena that have diffused from one neighborhood to another in historical sequence following paths of functional interdependence, and subject to the barriers of racial constraint, neighborhoods are open rather than closed systems. A different spatial systems framework is then necessary to specify the web of relationships within such observational units, as open systems, are set. Spatial autocorrelation is said to exist. The statistical analysis used must then be appropriate to that situation, and an alternative theoretical stance must be sought appropriate to that alternative situation.

This involves recognizing explicitly in the model that spatial systems display territoriality (areas of organization), formed with respect to points of focus that are structured hierarchically by dominance and subordination characteristics, through operation of distance attenuation mechanisms and boundary effects affecting interaction patterns over regular lines and channels of movement and communication. It also involves assuming some theoretical stance about the elements holding individuals together in systems – and in social philosophy there have been many variants of at least three differing stances (Bell 1971

1. *functional interdependence theories,* including (a) theories of exchange relations and market mechanisms (Smith) and (b) ideas of hierarchies of stratification based on technical competence (St Simon);
2. *value-integrative theories,* of which there are at least four – (a) rule by myth (Plato), (b) society as sacred (Burke), (c) society as a moral center (Durkheim), and (d) society defined by ends, traditional or concensual (Locke);
3. *domination theories,* (a) by traditional or irrational forces (Weber), (b) by the sovereign or the state (Hobbes), and (c) by class (Marx).

And what then of the individual observations? It follows from the above that they can only be understood in a relativistic sense, with respect to the entire structure of the system of which they are parts. If the system changes, so does the relative position of the individuals, which thus have no absolute and independent existence of their own yet which, in combination with all other individuals, define the system of which they are part. How, then, can one proceed at this juncture?

Towards process metageography

It was Emerson who said that "a foolish consistency is the hobgoblin of little men", and clearly what is called for in the above is not consistent application of any past methodology, but some new paradigm — a new conceptual statement that by its nature admittedly cannot be disproved, yet is open-ended enough to generate new rounds of research activity, serving for this activity as a basic statement of belief.

David Harvey (1969), among others, has argued that one step in producing such a metamorphosis of geography would seem to be that of examining interactions between temporal process and spatial form. To move beyond this suggestion, however (which is what I seek to do in what follows) a conceptual chasm must be bridged, for form can never be absolute. Not only is the "reality" of any element within a system relative to the entire system of elements; it is also time-relative. To seek any fixed thing is to deal in false imagination, therefore, for all phenomenal existence is immediately also seen to be transitory when the dimension of time is added. No particular thing is "real" in any absolute sense; it is passing into something else at every moment. Every individual, for example, is a progressively ageing, temporarily-organized "bundle" of energy flows faced with ultimate disintegration.

To be sure, the search for absolutes of form in some geometric sense is understandable. We perceive the world through screens composed of ideas, and the idea-systems are limited by a language oriented to classifying objects, naming things, and hence codifying their "reality". Yet what is needed to advance our science is conditional thought that recognizes the relativity of existence and the relative truth of perceptions. Indeed, what is needed is the initiation of a more continuous intellectual process in geography that recognizes that every system and every interpretation needs reassessment in the light of a more complete system.

As a contribution to such a process, I propose that geographic explanation be viewed as dealing with the antecedents and consequences of environmental and locational decision-making in which man, as the prime actor, is viewed as "an information-processing, decision-making, cybernetic machine whose value systems are built up by feedback processes from his environment. These feedback processes are built into the most primitive forms of life, and they form a continuous spectrum all the way back through prehistory and to times when no life existed. Throughout this whole development of man's history, coming up through biological evolution and extending into cultural evolution, the essential message is one in which disorder, or randomness, is used to generate novelty, and natural selection then generates order" (Potter 1971, 36).

This view implies looking at the world as a complex living system in which individuals, social groups and institutions are dynamically interrelated actors involved in continuing processes of decision-making. The nature, purpose and meaning of any actor and action can only be understood in relation to a field of forces involving other actors and actions. Many actions appear random, but disorder is ordered through responses to their consequences, which may reinforce or change nature, purpose and meaning by changing relations in the system. The behavior of the actors in such an eventuality contributes to equilibrium processes. Another way that the structure is maintained is by making homeostatic adjustments to disturbances; random trends away from system integrity are suppressed by negative feedback. Yet other actions can produce evolutionary shifts in structure by engendering or supporting morphogenetic processes involving gradual growth and change and increasing levels of organization. Finally, and more radical, are those actions that result in revolutionary transformations of structure. In each case, of course, decisions are made in the relational context of perceived organization and structure, and processes set in motion by actions therefore reaffirm or reform the intrinsic self-organization of the system amidst the apparent disorder of myriad decisions and actions.

What, then, is proposed is a view of the world from the vantage of *process metageography*. By metageography is meant that part of geographic speculation dealing with the principles lying behind perceptions of reality, and transcending them, including such concepts as essence, cause and identity.

Process metaphysics, the basis of process metageography, has been present in Western thought at least from the time of the Greeks. Whereas Democritus had argued that nature comprised a set of objects (atoms) in the void, Heraclitus said that all was flow (fire) [Platt 1970]. The fundamental idea of process metaphysics is that the universe should not be regarded as made up of objects or things, but of a complex hierarchy of smaller and large flow patterns (i.e. processes) set within systems of even larger scales in which the "things" are self-maintaining or self-repeating features of the flow with a certain invariance, even though matter, energy and information are continually flowing through them. The shape of a waterfall, the flame of a candle, or the shapes of clouds, which have a certain constancy even though masses of moist air are flowing through them and continually condensing and evaporating, would be examples. Similarly, in urban geography, the neighborhoods of a city retain their characteristics only because the same kind of people move in and out; such self-maintaining flows preserve the social geography of the city. Indeed, if the same people remain, the social geography changes because these people inevitably do so.

In such a flow-picture, the steady-state patterns or "objects" — or steady-state organisms or observers — can only be understood in a holistic relationship to their "environment", with fields of flow extending outward indefinitely to the next such stable concentration of energy, and the next. Likewise, the environment only takes on stable form and meaning and points of reference through the "objects" which it sustains. In this sense, electrons and the fundamental particles of physics, may be regarded as patterns or perhaps something like knots being knotted or unknotted in a field of flow that extends throughout the universe, or, as Toulmin (1962, 301) puts it, as "tidal waves in a sea of field-energy". These fundamental patterns are assembled, of course, into larger but less stable patterns, such as chemical molecules, living cells, organisms, brains and social networks and nations. The higher structures are built up in a hierarchy — in an "architecture of complexity" comprising complex systems and subsystems, but the emphasis in process metaphysics is not on the static structures of complexity, like the parts of a watch, but on a flow hierarchy, like the system of vortices, say, below a waterfall — that is, on structures that are self-maintaining or self-repeating with a certain invariance, even though matter, energy and information are continually flowing through them. Importantly for what we will develop later, such flow systems can undergo sudden transitions to new self-maintaining arrangements which will in turn be stable for a long time. Vortex patterns in a stream can be restructured in this way by a very slight motion of a stick or a rock, and they are sometimes unstable, flipping back and forth rapidly from one pattern to a quite different one.

As McLoughlin and Webster point out in their recent review of cybernetic and general system research (1970), deeper under-standing of these organizational mechanisms should lead investi-gators to the sources of intrinsic order and the nature of disorder, and, in human systems where decision-making plays a central role should also provide insights into the most appropriate strategies for deliberate or extrinsically applied controls designed to produce restructuring conformal to social goals.

The central importance of purposive behavior

At the core of the proposed system — the reader will recognize it to be a logical extension of my previous attempts to develop spatial field theories — is the notion of environmental and locational decision-making as the basis of the actions and processes maintaining geographic order or producing geographic change. I view a process as a repetitive or sequential cumulation of individual actions.

Such an extension is consistent with recent calls for an increased behavioral content of the discipline, and it implies a focus on directive or purposive behavior which at its highest level, brings us into the world of policy and social action. Alker (in press) has pointed out that all complex living or quasi-living systems are endowed with varying degrees of self-organizing, self-maintaining, self-reproducing and self-transforming capabilities. In such systems all actors, individuals and social units face, and to some extent respond to, formidable problems of autonomously organizing the appropriate instrumentalities, both symbolic and material, in the face of an uncertain and frequently probing environment. The idea of purposive behavior embraces all actual or apparent, conscious or unconscious goal-directed activity related to self-organization, self-maintenance, self-reproduction and self-transformation. The following are illustrative of the wide variety of such behaviors [Alker]:

1. *genetic mutation and selective survival* of those systems nicely or poorly matching their environments;
2. *bisexuality and heterozygosity*, which have generally enhanced the capabilities of a species in the "natural selection" process;
3. *blind trial and error problem-solving;*
4. *learning,* or the retention of adaptive (or nonadaptive) response patterns for subsequent use, thus speeding up the trial and error process for familiar problem situations;
5. *anticipation (or perception),* cognitive or visual exploration of potential behavioral alternatives, substituting for overt exploration;
6. *observational learning,* characteristic of social animals who learn from observing others' explorative attempts;
7. *imitation,* the acquisition of a model for behavior by perception of another's behavior;
8. *linguistic instruction* about the nature of the environment and of "correct" responses to it;
9. *cognitive rehearsal (or thought),* the symbolic exploration of potential behaviors vis-a-vis a learned model of the environment;
10. *social planning (or decision making),* wherein the knowledge processes noted above are combined across various individuals in ways superior (or inferior) to those of any single person.

Potter (1970) views the "purposive" organization in many of these processes as one of "directed chance" in which, first, there is the control of homeostatic regulation as described by Walter B. Cannon's six postulates (Potter, 119):

1. In an open system, compounded of unstable material and subjected continually to disturbing conditions, constancy is in itself evidence that agencies are acting, or ready to act, to maintain this constancy.
2. If a state remains steady it does so because any tendency towards change is automatically met by increased effectiveness of the factor or factors which resist the change.
3. Any factor which operates to maintain a steady state by action in one direction does not also act at the same point in the opposite direction.

4. Homeostatic agents, antagonistic in one region, may be co-operative in another region.

5. The regulatory system which determines a homeostatic state may comprise a number of cooperating factors brought into action at the same time or successively.

6. When a factor is known which can shift a homeostatic state in one direction it is reasonable to look for automatic control of that factor or for a factor or factors having an opposing effect.

Failing such homeostatis, he argues, secondarily, for the possibility of the direction being towards adaptive change of three kinds (p. 124).

1. *Evolutionary adaptation,* involving populations over a period of many generations, the process by which natural selection acts on a population of nonidentical individuals and selects those whose heredity is best suited to reproduce in the given environment. This kind of adaptation cannot foresee future environments and often leads to extinction but need not do so.

2. *Physiological adaptation,* the process that each of us as individuals are capable of over time periods that range from minutes to weeks to years, including the orchestration of a symphony of individual organs, and adaptation at the cellular level, wherein enzyme activities and enzyme amounts wax and wane in response to need.

3. *Cultural adaptation,* the psychological counterpart of physiological adaptation in the individual, but analogous to evolutionary adaptation when populations are involved.

Particularly in the latter human contexts, the study of "purposive" activity is not to search for invariant causal connections, but to look for the available alternatives and to ask why the agent actualized one rather than another. The explanation of a choice between alternatives is a matter of making clear what the agent's criterion was and why he made use of this criterion rather than another and to explain why the use of this criterion appears rational to those who invoke it. Clearly the problems and alternatives perceived, the decisional criteria invoked in a particular context, and the more or less illusory sense of rationality occur *because of or in spite of* various reasons, goals, intentions and causes, and their existence should be built into any framework for analysis and study.

A behavioral model of spatial process

It is the preceding considerations that led to the development of the paradigm of environmental and locational decision-making and planning that as modelled in fig.1.1. In the model all decision-making is viewed as taking place in a locational and environmental context — an ecosystem that is a functioning interaction system of living organisms, including man, and their effective physical, biological and cultural environments. This ecosystem is a product of interacting natural and cultural processes. However, one long feedback loop shows both natural and cultural processes to be, in turn, affected or created by spatial processes composed of prior decision-making sequences made in prior environmental contexts, and driven by combination of biological needs (survival, maintenance, reproduction) and cultural drives (such as the need for achievement built into the central nervous system of individuals in societies displaying economic and technological progress through the altitudes and pressures of the culture (McClelland 1961).

In the system, decisions relate to ecosystem through intervening perceptual filters that bias the feelings of need for action and of capacity to effect change, while shaping and coloring the actor's mental maps of the ecosystem and his action space within it. Perceptions, in turn, are a product of biological needs and constraints, natural endowments, and the actor's world view and cognitive structure, based on the values of his culture and the roles, expectations and aspirations imposed on its members, together with the fruits of learning based on experience with the results of prior decision-making and action. Indeed, beliefs and perceptions may be among the most critical elements because what men believe determines what men do. For example, beliefs about the world determine the planning style selected, and at least four such styles are suggested in Table 1.1. The most common is simply *ameliorative problem-solving* — the natural tendency to do nothing until undesirable dysfunctions are perceived to exist in the system in sufficient amounts to demand action. This orientation contrasts markedly with

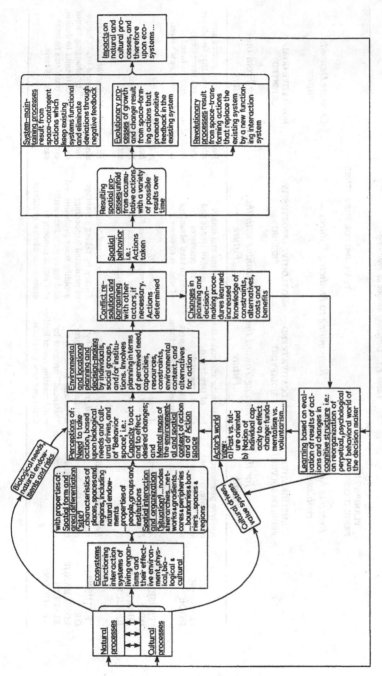

Figure 1.1 Behavior and Process in Ecosystems

Table 1.1. Differing Modes of Planning

	PLANNING FOR PRESENT CONCERNS	PLANNING FOR THE FUTURE		
	Reacting to Past Problems	*Responding to Predicted Futures*		*Creating Desired Future*
	AMELIORATIVE PROBLEM-SOLVING	ALLOCATIVE TREND-MODIFYING	EXPLOITIVE OPPORTUNITY-SEEKING	NORMATIVE GOAL-ORIENTED
	Planning for the Present	*Planning Toward the Future*	*Planning with the Future*	*Planning From the Future*
PLANNING MODE	Analyze problems, design interventions, allocate resources accordingly.	Determine and make the *best* of trends and allocate resources in accordance with desires to promote or alter them.	Determine and make the *most* of trends and allocate resources so as to take advantage of what is to come.	Decide on the *future desired* and allocate resources so that trends are changed or created accordingly. Desired future may be based on present, predicted or new values.
"PRESENT" OR SHORT RANGE RESULTS	*Ameliorate Present Problems*	*A Sense of Hope* New Allocations shift activities	*A Sense of Triumphing Over Fate* New Allocations shift activities	*A Sense of Creating Destiny* New allocations shift activities
"FUTURE" OR LONG RANGE RESULTS OF ACTIONS	*Haphazardly Modify the Future* by reducing the future burden and sequelae of present problems.	*Gently Balance and Modify the Future* by avoiding predicted problems and achieving a "balanced" progress to avoid creating major bottlenecks and new problems.	*Unbalance and Modify the Future* by taking advantage of predicted happenings, avoiding some problems and cashing in on others without major concern for emergence of new problems.	*Extensively Modify the Future* by aiming for what could be. "Change the predictions" by changing values or goals, match outcomes to desires, avoid or change problems to ones easier to handle or tolerate.

future-oriented planning to achieve goals that are either latent in cultural values or explicitly recognized and stated. Use of one planning mode as opposed to another is a product of the decision-maker's world view, whether past- or future-oriented, and his concept of the role of self in change, ranging from fundamentalist to voluntarist.

Planned decisions only translate into action when conflicts with other actors have been resolved. Once conflict is resolved, though, the resulting actions taken give rise to locational and environmental behavior: movements, locations, relocations, resource use, etc. But these change the spatial system, so that if decision-making is a biologically and culturally driven and perceptually biased creature of environment, it is also a creator of environment through impacts on natural and cultural processes resulting from spatial behavior, as is shown in the long feedback loop in fig. 1.1.

Individual actions are, of course, myriad, and it is useful to think of them as events which, in repetitive or cumulative sequences, contribute to spatial processes of one of three kinds:

1. *System-maintaining.* This involves repetitive process events that either keep the system functioning or which, in a cybernetic sense, seek to eliminate perceived dysfunctions and deviationist tendencies through negative feedback. It is such processes that maintain steady-state patterns.
2. *Evolutionary.* These events are those which, in cumulative morphogenetic sequences, produce growth and progressive change by amplifying positive feedback in the system.
3. *Revolutionary.* Such events set in motion sequences that transform the system by redefining its members, limits and styles and types of interactions.

These three kinds of spatial processes have also been called space-contingent, space-forming and space-transforming. An important distinction lies between the first, which involves those rhythmic repetitive events that characterize a given system and are essentially processes *in* history, like the daily ebb and flow of commuters, and the second and third which refer explicitly to change and to processes which *are* history, as when industry and people relocate. If the first provides complex systems with powerful self-organization that tends to suppress change, it is the latter which embed in such systems the capacity for self-transformation into new and different states.

Where the transformation involves revolutionary restructuring, it has been characterized by Platt as involving some "hierarchical jump" (1970). He points out that the universe seems to be made up of sets of systems, each contained within one somewhat bigger, and that self-generated jumps in a hierarchical organization have several common characteristics that stand out. One is that the jumps are always preceded and accompanied by "cognitive dissonance".

Thus, Kuhn (1962) describes in considerable detail the scientific dissonance that precedes scientific revolutions. First, there are accumulating bits of data that do not fit the old predictions; or rules of thumb in certain areas that seem to be justified only by odd assumptions. In the beginning, these difficulties are dismissed as trivial or as errors of measurement or crackpot arguments, but they do not go away, and they get more numerous. After a time, the confrontation with the old system comes to be recognized as fundamental, and various proposals for a reconciliation are brought forward. Then suddenly a simplification from some entirely different point of view makes big parts of the problem snap into new and clearer relationships. There is a collective sense of relief and achievement, even though a long period of working-out may lie ahead.

A second feature of self-generated hierarchical jumps is the overall character of the dissonance and of the later transformation. Thus, the Industrial Revolution turned out to be a revolution in attitudes, banking, commercial organization and city structure as well as in technology.

A third striking feature of hierarchical jumps is the suddenness of the restructuring when it arrives. The Russian Revolution shook the world in ten days, and the U.S. Constitution was hammered out in a few weeks.

A fourth characteristic is "simplification". In scientific advances, the direction of advance is always towards simpler and more general explanations. Whether or not such simplification is also present when spatial systems are restructured is certainly worthy of examination.

Perhaps the greatest bulk of geographical study has focussed in past years on those properties of the environment — spatial form and areal differentiation, spatial interaction and organization at particular points in time — that geographers have variously considered their proper domain, as is suggested in the environmental "box" in fig. 1.1. From this there has sometimes been the attempt to infer process. But to paraphrase A.N. Whitehead, process is the *becoming* of experience ... and experience has (therefore, in geography) been explained in a topsy-turvy fashion, the wrong end first. What we are proposing here therefore is a hierarchical jump in geographic explanation. A process view involves studying the continuous attempt to attain ends, with *becoming,* which derives its ultimate meaning from either repetitive sequences or change sequences of process-events.

The continuity of process *in* history resides in repetition or reiteration of spatial behavior — in collections of activities or events arranged in particular sequences in which actors repeatedly engage. Such sequences are often rhythmic on daily, weekly or seasonal bases, and are frequently legitimized by legally-formulated procedures or institutionalized practices which have the effect of defining and constraining illegitimate behavior. The resulting system-maintaining spatial processes retain longer-term steady-state geographical patterns.

Change processes, on the other hand, are always problematical in that they necessarily involve challenges to the organizing elements of legitimacy, formal procedures and institutionalized patterns of behavior. This implies that to identify change processes one must have a basis for sensing shifts — a known starting point in the reiterative patterns of existing processes. Equally, it is important to determine whether the shifts are based upon process-changes within the system, or systemic transformations of a more fundamental character producing hierarchical jumps in self organization.

Among the more pressing research priorities that flow from the paradigm then, are:

1. logical clarification of the variety of action modalities;
2. the formal treatment of decision and action sequences;
3. the further analysis of emergent properties, including systemic transformations. (Alker)

Concern with such issues should enable geographers to address more clearly a central problem facing societies today: the definition and possibly the conscious pursuit of desirable collective ends by citizens aware of the actual tendencies and capabilities of imperfectly organized collectivities and sensitive to individual needs, capabilities and limitations.

References

ALKER, H.R. Jr. (in press). Directive behavior: a desirable orientation for mathematical social science; *Revue Française de Sociologie.*

BELL, D. (1971) *Theories of Social Change*; (Russell Sage Foundation, New York).

BRUNN, S.T. and HOFFMAN, W.L. (1970) The spatial response of negroes and whites toward open housing: The Flint Referendum; *Annals of the Association of American Geographers* 60, 18-36.

CLIFF, A.D. and ORD, J.K. (1969a) The problem of spatial autocorrelation; In SCOTT, A.J. (Ed.), *Studies in Regional Science*; (Pion, London, p. 25-6).

GOULD P. (1970) Is *Statistix Inferens* the geographical name for a wild goose?; *Economic Geography* 46, 439-48.

HARVEY, D. (1969) *Explanation in Geography*; (Edward Arnold, London).

HILDRETH, R.E. and NAROLL, R. (1972) Galton's problem in cross-national studies; In BINDER, L. and VERBA, S. (Eds.), *Comparative Studies of Political Development*; (Princeton University Press).

KATZ, E. LEVIN, M.L. and HAMILTON, H. (1963) Traditions of research in the diffusion of innovations; *American Sociological Review* 28, 237-52.

KUHN, T.S. (1962) *The Structure of Scientific Revolutions*; (University of Chicago Press).

LOWIE, R.H. (1946) Evolution in cultural anthropology: A reply to Leslie White; *American Anthropologist* 48, 227-30.

McCLELLAND, D.C. (1961) *The Achieving Society*; (Van Nostrand, Princeton).

McLOUGHLIN, J. and WEBSTER, J.N. (1970) Cybernetic and general-system approaches to urban and regional research: A review of the literature; *Environment and Planning* 28, 369-408.

MIKESELL, M. (1969) The borderlands of geography as a social science; In SHARIF, M. and SHARIF, C. (Eds.), *Interdisciplinary Relationships in the Social Sciences*, (Aldine, Chicago), p. 277-48.

PLATT, J. (1970) Hierarchical growth; *Bulletin of the Atomic Scientists* (Nov.), 2-4, 46-8.
POTTER, Van R. (1971) *Bioethics*; (Prentice-Hall, Englewood Cliffs).
TOULMIN, S. and GOODFIELD, J. (1962) *The Architecture of Matter*; (Harper and Row, New York).
TYLOR, E. (1889) On a method of investigating the development of institutions applied to the laws of marriage and descent; *Journal of the Royal Anthropological Institute* 18, 272 ff.

2 · Between theory and metatheory*

DIETRICH BARTELS

*Today everyman wants his crisis. The word has lost its
disturbing sound. . . . Thus our universities too have been
living for some time with their crisis. There are, however,
crises of very varying type: those of growth, of transition
and of decline.*

W. Hofmann 1968, 9

Science as a process lives from the renewal of its standpoints; its
progress, it is true, also proceeds in lateral growth but predominantly
as a forward-thrusting reconstruction. It is a basic rule that the
relevant innovations seldom spring up at the top of the scientific
community but mostly at the base, where younger people, for the
most varied reasons, perhaps because of modified bases of credibility
or re-oriented value co-ordinates, fail to accept inherited trains of
thought and subject them to unfamiliar critical examination. This
has advantages and disadvantages. On the one hand, continuity is
better preserved in the age-structured hierarchy of a discipline's
organisation wherein innovations come up against an institution-
alised resistance which acts as a testing filter. On the other hand, the
considerable delays which can occur in the acceptance of these new
ideas often results in disproportionate conflicts which cause mere
changes of a leading paradigm to assume the temporary character of
a revolution (Kuhn 1962).

Since scientific research, which may be looked upon as a system
of labour-dividing functional interrelationships, necessarily requires
an organizational structure in which experience of life is indispensable or at least very desirable, and yet since at the same time the
stream of innovation from below is gathering volume and impetus,
this conflict is currently becoming more acute. In order to overcome

---

* A first German version was published in *Geographische Rundschau* 22,
1970, p. 451-7.

this conflict it is necessary, as a first step, to recognise in this sharpening of the situation not only the continuing tendency towards dismantling the inherited structures of authority but also the basic psychological fact that fundamental innovations in the vision of the world, cannot be, in a simple manner, intellectually accepted by all scientists. This is especially true when innovations make the results of an individual's research appear to have been robbed of their relevance. Thus many innovations must have their diffusion delayed, and most 'grow through' as part of a transfer of power between generations. Plurality of points of view at any one time is thus unavoidable in a discipline and our reaction to this should not be one of dismay but a recognition of the need for a corresponding expansion of democratic forms of pluralistic co-existence in science which accepts these situations of conflict between different 'statements of truth'.

There are, naturally, two important limits to such a pluralism within a discipline:

(a) The internal differences must not be greater than those which exist with neighbouring disciplines, otherwise the whole reason for the institutional organisation of the scholarly discipline is lost. It seems as if we are currently finding ourselves very close to this situation when, for example, one thinks of the relationships between physical geography and the other natural sciences in the field of natural ecology, or of the convergence of interests between human geography and other social sciences. The oft-quoted 'bridging' function between subject areas which is supposed to be the special task of geography is currently creating severe methodological problems for us.

(b) The reputation of geography as a practical science can only be secured by its possession of a clear-defined public image. There has thus got to be a certain general consensus not only regarding the possibilities of using geography in the direct solution of acute social problems but also regarding its importance as a school subject, the content of which must be guided by geography in the universities.

It is therefore essential for geography, whose long traditions tend to be those of particularly well-developed pluralism, to gather together and to re-formulate its basic concepts; particularly those which create or ensure a certain disciplinary identity and present a clear profile to society. The choice of these basic concepts depends on what practical tasks are allocated to the discipline and its place in the school curriculum. A precondition for such a new formulation is the clarification of current tendencies in our discipline, such as I attempt below.

If one seeks to pass beyond individual matters of research in geography (whether new or reinterpreted) and to consider innovations in our basic scientific concepts, one gains the strong impression that, as in other disciplines, there is in certain directions a growing *rationality* in our attempts to structure the basic concepts of geography and in our attitudes towards them. The term 'rationality' itself has recently been vigorously transferring itself from technology and economics to the organisation and operations of science. It is possible to identify a number of different innovation waves in methodological attitudes and theoretical problem areas, which are actually or potentially attributable to this process, however crude the interpretation assumed by supporters of such methodologies. In the paragraphs which follow an attempt is made to describe this growth or rationality within the sequence of new standpoints which are characteristic of developments within geography. In this way the *plurality* of temporally-overlapping basic orientations which dominate the contemporary geographical situation may perhaps become clearer.

Instrumental rationality

The ideals of precision and exactness, of the neutralisation of thought against subjective peripheral influences, and of the control of complex problems and large masses of information are those which first can be employed in judging the rationality of the methods and the evidence of results of research. Mathematization, quantification and formalization are the most pronounced features of the advances involved in this *particular* aspect of rationality. Even in aspects of scholarship which have previously seemed not particularly amenable to such methods axiomatic-deductive thought and statistical procedures are now being pursued. Measurement, model-building, programming, stochastic processes and systems analysis, forecasting and control systems are becoming operational concepts which are of growing importance to the development of science, technology and economics, and therefore to civilisation as a whole. The results of this appear to exercise increasing influences over the life styles of the population in an industrial society. It is not unreasonable to equate these aspects of modern existence with the very idea of rationality. In science they seem to stand for a maximum of objectivity, neutrality and value-freedom (attributes which are here either taken as self-explanatory, or as being in the main preconditions, and clearly separated from the political-moral aspect of value-based interest dimension of life). Science will naturally maintain these characteristics in their core meaning, since in a highgrade society based on the division of mental labour the two tasks of cognition and of decision cannot be recombined arbitrarily.

In geography, this instrumental rationality broke through at the beginning of the 1950's as the so-called 'quantitative revolution' and did so first, significantly, in the U.S.A., the leading scholars of which possessed a clear technical-pragmatic orientation supported by analytical, positivistic theory of science. In the 1960's these relevant methods and values spread from the U.S.A. into European geography.

It should be pointed out that quantitative work with measurement and numbers is not, in essence, a complete innovation. The attempts at measurable comparison of phenomena of physical geography beginning in the second half of the 19th century or the Associations of that time for 'geography *and* statistics' in Germany can be seen as forerunners of this development. Indeed, Haggett (1965, 15ff) and Lukermann (1965), for example, refer to such traditional quantitative efforts. More appropriate than the currently-widespread word 'quantitative' in characterising these new trends is the use of the term 'theoretical geography' in the sense in which Bunge (1962) introduced it. The application of theories, that is, conceptual models of reality, as inherent structural elements of all discursive human thought is, of course, fundamental. What is particularly important, however, is the growing volume of *deliberate,* controlled and controllable construction and use of such theories as a means of achieving a partial explanation of the experience of reality. Theories which, on the one hand, attain an unusually high level of abstraction partly with the help of formalized languages and which, on the other hand, precisely because of this, require the evaluation of large amounts of data in order to carry their empirical verification further than simple assertations of plausibility. In 1933 Christaller was an isolated example of a model-theoretical researcher in geography, but with our rapidly-expanding capability of constructing theoretical models in the late 1950's and early 1960's a growing number of formalized theories were applied in geography which were found to have surprising analogies and extensions beyond their special areas of application. In addition, there developed a reservoir of complex statistical techniques the successful applications of which led to a feeling of victory for this first wave of rationality. This euphoria was restrained by warnings from some innovators, for example by Burton (1963) and Curry (1967), that the situation which had been brought about was only one stage in a continuing process of development.

Differences in the understanding of theory

Particular emphasis, therefore, within this instrumental rationality is thus laid upon the concept of *theory*. Theories here are defined as larger, generally-applicable systems of a cognitive character which show, firstly, an inner consistency between their elements and/or, secondly, are formulated operationally in the language of observation. The two features involved in this definition have commonly been considered in isolation, and this has led to imbalance. There has developed, on the one hand, a deductive 'model platonism' (Albert 1968) linguistically immunised against empirical experience, as for example in the beginnings of regional science, and, on the other hand, a set of empirical concepts perfect only in their descriptive power, as for example in measures of centrography. Such imbalance has caused important disputes *within* the camp of instrumental rationality, but these have been gradually overcome.

In contrast to this concept of theory as a set of empirical-operational analytical linkages with an explicit hypothetical basis and with predictive power, stood and still stands an older and completely-differently-oriented idea, whose idealistic root is obvious, wherever this idea seeks to inquire into the 'true nature' of the perceived object, and especially into the necessary and essential attributes which, it is believed, are already *a priori* pre-registered in the phenomena itself. The 'logical system of geography' has been postulated as the stepwise integration of reality into the geographical entity of *'Landschaft'* in which one seeks to uncover the individual regional 'harmonies' and 'dominants'. In this sense theory is understood as the really existing and intellectually tangible background of our world, as exemplified by such terms as the 'totality of the landscape' and the 'classification of regional substance'. The definition of these terms present no particular problem to these theorists so long as certain primary postulates are agreed on, and in particular that the external appearance of the world is a directly-visible expression of an underlying unity of nature and spirit. The existential experience of the synthesised form of the object under study, moulded individually and historically, for which the consensus of researchers was taken as indicative of its reality (as was indeed broadly the case), so long as the plausibility of the suggested explanations which were deduced from this visible picture adequately confirmed this reality and so long as the manysidedness of heterogenous explanatory backgrounds was taken as a virtue and as a genuine basis for justification.

Theory appears here not as an instrumental part of the basis of an analytical type of research which is finally oriented towards the technical *manipulation* of the environment, but as the extension of the idealized foundations of a reflexive *interpretation* of the world [1]. Manifestations of this understanding of 'theory', often called essentialistic, can still be found today, as for example in the concepts of 'regional personality' or the 'objectivised spirit in the cultural landscape' ("objektivierter Geist in der Kulturlandschaft"), the 'real economic region' ("realer Wirtschaftsraum") or the substantiated 'landscape process field' ("landschaftliches Prozessfeld"). We recognise in such conceptual ideas a high level of integrative forces, the so-called synthesising capacity, which resulted in the classic regional monographs, with their multitude of conceptual viewpoints, breadth of subject matter and attempts at overall regional interpretations.

Many disputes over geography's understanding of its own position were strengthened into apparently insoluble dichotomies by these varying attitudes towards the conception of theory. On the one hand there was speculative causal research into the totality of an individual geographical example, on the other there was the stochastic testing of correlations and generalized functional relationships. The unified vision was set against systems analysis. The description of individual historical phenomena in the examination and verification of seemingly securely-based interpretations of reality was thus contrasted with generalizations regarding components of reality having the aim of the construction of instrumental theories of at least medium range (Merton) or of a quasi-law nature in the sense used by Popper. In this dispute, superiority felt by the supporters of instrumental rationality was often matched by a desperation on the part of the 'essentialists' which led often the latter to the use of arguments inferior to those which were actually available for the systematic defence of their fundamental position. *Landschaft* geographers and regional geographers – insofar as they did not perceive a compensatory irony in the fact that quantitative methods could often only confirm previous trivial qualitative views – often remained indignant and unappreciative, seeing no competition with their own academic goals, since the theoretical discussion of the other camp seemed to have been shifted to a completely different level from their own. However, the view was occasionally held that it might be possible to harness the new instrumental rationality in order to attain the goal of total synthesis of all geographical phenomena. Examples include the landscape-axiom theory of Neef (1956), Bobek's (1962) proposal to include newer aspects of the process of research into cultural landscape analysis, and Uhlig's (1970) draft for a self-contained uniform system of geography. In all this, however, the maintenance of the essential difference between the two methodologies remains important. In the older school theory is treated as identical with, or at least based on, a sensitive value-background of personal experience which was ontologically over-emphasized as the total horizon of life. Approaches of this kind have been beyond the instrumental rationalist as long as he has simply operationalized some parts of the traditional 'questions of the nature of things' in geography (problems of natural determinism, universal regional structure, etc.). An example can be taken from the first generation of North American 'revolutionaries' who took little notice of the development taking place at the same time in Europe, especially in Germany, towards new questions in social geography, in the form which Ruehl, Waibel, Busch-Zantner, Winkler, Pfeifer and Bobek had successively developed between 1930 and 1948.

Thus instrumental rationality has neglected the *creative relationships* which always contain the general cognitive patterns as well as value elements of society. Its practitioners have sought to make up for the lack of these creative relationships by the use of simulation procedures leading to the automatic production of hypotheses. In this endeavour the individual researcher draws his satisfaction only from the positive *verification of relationships,* that is, from the possibility of empirical confirmation or rejection of given hypotheses. The 'essentialist' as geographer, on the other hand, has to a large extent dealt with his creative relationships in terms of the paradigms of his personal landscape experience and of his interpretation of the structure of space. From these he drew his scientific conviction and confidence, without, however, examining these creative relationships more clearly in terms of their nature, justification and explanatory power. That is to say, he neglected what I would call the metatheoretical aspect of an approach of research.

The limits of the 'freedom of values' concept

As a result of the above comparison the deficiency in both positions becomes clear, namely the absence of the testing of research attitudes from the point of view of the theory of cognition. This has been recently acknowledged in a *second* wave of rationality in a number of sciences, including geography. Rationality has now come to include a problem-consciousness with respect to current basic assumptions, as opposed to the uncritical acceptance of self-generated or adopted ideological viewpoints regarding the empirical world. This is revealing as illusory many of those scientific 'explanations' which have hitherto appeared plausible, and is posing the more fundamental question regarding the definition of 'satisfactory explanations', as the aim of research has often been defined.

The critics have come up against the 'unexplainable residual' or 'metaphysical component' in the most empirical derivations. They have encountered the need to base every rational theory on what is, in the last resort, an irrational 'prior understanding' of the given situation as a metatheoretical background which has, at least until now, been stubbornly excluded from scientific debate. Such a realization of the importance of pre-scientific 'world perspectives' (Hempel) and their key terms, fundamental categories in the personal experience of reality, as forming the precondition of so-called objective and value-free research, marks the second stage of rationality.

Two attitudes can develop from the realization. The first is a *relativism* of scientific attitude which draws support from a majority of workers within the discipline, leaving the metatheoretical problem unexplained as 'a spanner in the works'. The other is a *nominalism* which attempts to systematize various world perspectives as basic stimuli to concrete research, to analyse their content and to test them metatheoretically as to their functions and their standpoint. In this second stage of rationality the possibility has been revealed of reducing the abundance of individual geographical research theories (often differentiated or simplified by ceteris paribus-assumptions), to certain paradigmatic basic horizons (e.g. theories of water circulation, of exogenous relief formation, of vegetation equilibrium, or of deterministic anthropogeography) on the basis of a preoccupation with natural ecology in the context of a physiognomically-scaled model of 'geofactors'. To take another case, central place theory, industrial location theory and the range-of-a-good behaviour approaches, all initially constructed on the assumption of ideal 'economic man' (models which now show increasing disaggregation in the direction of the bearer of individual decisions, the process of which still needs to be more analysed), have since 1950 been relativized as special cases of a more general basic social science orientation involving man as an interactor and his norms (in the sense of the conceptual development, stemming from work from Weber to Parsons), which still requires further development. These and other theories on which geography has concentrated are now being examined — in a manner supposed to be objective — for their unavoidable *pre-perception* of reality and the elements of this pre-perception.

Metatheoretical nominalism has made a number of interesting discoveries. For example, in the discipline of geography several basic tendencies now exist alongside one another which are impossible to integrate today since they differ from one another in every respect from their basic norms down to their subsidiary techniques. Again, it is recognised that such world perspectives and their derivatives, together with the framework of basic concepts, can only be poorly-united by the language of observation, but that on the other hand they usually develop a high measure of self-explanation and evidence. This feature transcends mere axioms, attains the ontological certainty of 'real truth' and is expressed in appropriate conceptual categories (e.g. substance, organism, budget, entropy, landscape, actor's decision, society, etc.), Thus Habermas (1968) speaks of 'transcendental points of view' and Topitsch (1965) of the 'metatheoretical aspects' which, through their critique of such categories, move more strongly into the interest of a rational theory of science.

This metatheoretical analysis of the character of basic concepts with their essentialistic impact brings among other things some relief to the hard-pressed defenders of *Landschaft,* for, as in a critical manner Hard (1969 and 1970) in particular has pointed out, it can be seen that the German theory of *Landschaft* possessed all the characteristics of a respectable basic theory and presents the best paradigm in geography to develop the appropriate metatheoretical insights. It was, in fact, not in any way the work of dilettantes or simple enthusiasts, as some have suggested, but the equivalent of a basic scientific world experience which was at least as well founded in transcendental ideas as many contemporary theories in, for example, the study of society. It is only that its plausibility and rationale, the product of classical German idealism, has been lost, and this has resulted in the lack of understanding of it exhibited by the younger generation.

Metatheoretical nominalism can, therefore, be seen as the *second* feature of the rational movement, as a realization of the plurality of existing basic-theoretical starting points and their tendency (however latent) towards ontologization and the assumption of an absolute 'nature' of their respective subjects. In a manner of speaking, the discovery has been made of those 'spectacles' necessary to research of which Toulmin (1968, 124) says that there is only one possibility of seeing them oneself, one must take them off! At the same time, however, one's intuitive process is interrupted until they are put on again — if that is possible [2]. The transition from the still-dominant methodology of instrumental rationality to the deeper under-standing of theory is shown clearly in the English-language method-ological literature. For example, the valuable textbooks by Haggett (1965) and by Cole and King (1968) treat geographical theories predominantly according to the formal criteria of their structure [3], whereas in other authors, above all in Harvey's (1969) fundamental introduction to methodology, the forces of basic theory are slowly being brought into play [4].

Metatheoretical criticism

A *third* and more penetrating step in the development of rationality in science pre-supposes the necessity of such 'spectacles' and their plurality, whilst also avoiding the attributes essential for the characterization of particular basic tenets. It enquires further into the motivations for the choice of a particular pair of Toulmin's spectacles which provide a plausible framework, and into the value judgements which have led to the adaptation of a particular world perspective. To express it in another way: the design of a basic theory must hold out some promise to the person who has formulated or adopted it, if he is going to let it influence his research. What are the norms which determined his decision? — Rationality at this stage takes account of the way in which the basic ideas of scientific activity and their selection have come about historically. Above all, it shows an historical-critical understanding of the relative positions of various geographical perspectives. In this way, it is possible to differentiate between the following two points of departure which are partly distinct and partly linked.

On the one hand we have seen revealed the semantic, linguistico-logical implications of basic theoretical viewpoints. In geography these emerged as a result of the analytical overcoming of the objective-idealistic goals of synthesis, as for example in the doctrine of *Landschaft*. Examination of the entire speech-associative context of the higher educational world of the 19th century, to which the distinctive vocabulary of geography used by our forebears inevitably belongs and which we can only escape by very conscious steps, has revealed the complexity of derivation of our prevailing basic concepts and the dependence of even our cognitive pattern on the respective communicative reference groups. It has, in fact, 'unmasked' our basic ideas and approaches from their linguistico-cultural constraints. On the other hand, we have the development of a *socially-critical* attitude towards science and its value premises as a partial phenomenon of social reality, which is indeed to formulate and explain 'rationally and without pre-conditions', but whose analysis is nevertheless itself tied to socially-determined preconceptions by a hermeneutic circle, in which science as a function of society is bound into its total system both as a carrier and as something carried — as is indeed postulated by our metatheoretical model of the process of cognition.

Linguistic analysis and social criticism as illuminators of the role of science and its chosen paradigms, and as providing a rationale for the contents and paradigms of research, become important components in the foundation of a discipline through the theory of science. They lead, as the third wave of growing rationality, to the identification of a feedback cycle which can be documented at various levels and which appears to be an irrevocable condition of research activity. A much-quoted aphorism by Wittgenstein (1963b, 114) is that "We were imprisoned by a *picture*. And we were unable to escape from it, because it lay within our language and that seemed to repeat itself only to inexorably bring it before us again." And also Habermas (1968, 19) maintained ". . . radical cognitive criticism is only possible as social theory".

The planning of research as a necessity

In another context we find, however, that Wittgenstein has given a different, existentialistically-striking impression when he wrote: "We feel that, even if all the possible scientific questions have been answered, our problems of life have hardly been touched upon." Thus, in the content of modern social criticism, we find a further dimension: the proximity and intermingling of 'cognition and interest', of passive experience and active engagement, that is, the necessity of critical *formation* of science as a task of society and certainly above all as a task of the scientist himself.

The efforts which have been made in geography in the direction of *relevant group decisions,* including those initiated by the associations of geography school teachers are, in my opinion, insofar as they build upon the stages sketched out above, a *fourth* consequent wave in the sequential model of growing rationality which I have attempted to outline. Its progress is a matter of great urgency, if only because the third step has led to reactions which are only partly rational [5] and which can be identified as 'emotionally-determined evasive action'. On the one hand there is the resignation in a linguistically-comprehended hermeneutic position, and on the other a hasty commitment to social engagement as a way of life which draws its justification from a dialectic standpoint, – although the support given by the latter in recent years to the breakthrough of rational creative efforts in society and scholarship is unmistakable. It is clear that it is high time that critical rationality at this fourth stage is applied to conscious shaping of scientific life. Hofmann (1968, 7) is of the opinion that "we have reached a situation in which the unsolved problems of the social order rebound onto the productive powers of society" and to these productive powers which call for such decisions the universities and their disciplines belong in pride of place. In this situation we cannot, of course, expect either that the deeply-penetrating cognitive criticism by linguistic analysis or social theory as a continuing process will of itself lead to new paradigms of scientific activity, or that it will absolve us from the decisions which have to be taken *now* about the functionally-sound contemporary basis of our discipline, however important its continuing aid to orientation may be in the future. To give a crude comparison: intensive long-term market and development research in a car factory can never render superfluous the maintenance of the current assembly line, even though each research advance shows it to be always slightly suboptimal [6].

We are, therefore, not excused from the necessity of making the constructive decisions currently required of us, namely: what basic ideas and theories are to form the basis of our subject and give it direction. Indeed, the confrontation becomes sharper inasmuch as the necessity for this decision – and its tentative nature – becomes clearer with growing rationality. The metatheoretical support for it may lie in the erection of certain minimum criteria for the relevance of the paradigms to be chosen, so that the collective decision of the planning of research will not become a step backward into a positivistic ignoring of the basic problems in favour of 'assembly line production' at any price.

Dispensing for the moment with the question of whether it is possible to make them operational, I would like experimentally to identify three dimensions from the formulation of such 'relevance' criteria. They doubtless need not be independent of one another and could perhaps be alternatives. Basic geographical concepts, for example our favourite 'human interaction in space', should:

1. Be of *social relevance* in the sense that they must be directed at least in the medium term [7] towards the solution of problems of social development which are considered to be acute and pressing. This holds good quite independently of the fact that only such an orientation will attract research funds to our science and that only such an orientation will allow researchers to enter into the 'functional elite' (Scheuch), the only rank-order of a classless democracy.
2. Be of *didactic relevance;* that is, it should be possible to gain transferable insights and skills from its deductions, of such a nature that they are of service in our control of the environment in a way which is relevant to individual person's way of life. This is particularly important as long as geography remains a major school subject.
3. Be last, but not least, *scientifically fruitful;* that is, it should relieve the researcher's attention of preliminary metatheoretical questions and to concentrate it, by stimulating intuition, on the 'normal' individual empirical problems of the particular object under study. In Toulmin's sense it should provide a well-fitting pair of spectacles whose functions one can normally forget, in order in a concrete sense to push forward the boundaries of the known against the unknown, or to improve on the reliability of present knowledge. Such required creativity presents probably hard conditions, previously little analysed, in terms of the basic concepts and their meaningfulness. For example, appropriately

dosed fronts of provocation as in the alienation of normal problems up to now and in the necessary communication with other disciplines and groups, and lastly also in taking the requirements of continuity of scholarship. Especially from this last condition, one is able to draw the conclusion that we cannot casually jeopardise the individual cognitive gains in theory made by our predecessors in the belief that we have ample possible choices to build a brand-new science from the ground up.

Looking at it in this way, we have to remain conscious of the historical growth of geographical science which is a development process in the sense of the above three criteria and, as is shown in fig. 2.1, whose stages operate simultaneously as components of plurality right up to the present day, even if they have very different emphasis (e.g. local or national schools).

Conclusion

Critical rationality of a fourth-stage type necessitates the full use of the growing possibilities of metatheoretical analysis and also the evaluation of the boundaries of its practicability, where it seeks to make its *tentative* but necessary decisions on the *relatively* optimal formation of a pluralistic situation and to its total social context (whose values need to be made more explicit) in a feedback situation: it is also the launching point for this reflection stage. It is in this sense that, in the last resort, one has to interpret that smooth phrase 'geography presents reality': in which one displays it with metatheoretical understanding in relation to the change in the social reality of our times.

Notes

1. In a similar sense some geographers of the German camp of social criticism, reject the application of one-sided positivistic scientific scales to the *Landschaft* geography which, after their inter-pretation, has served the 'dissemination of ideological learning' about countries and peoples.
2. Only the consequent use of a particular 'pair of spectacles' can evidently exercise the cultural relief function for the scientist

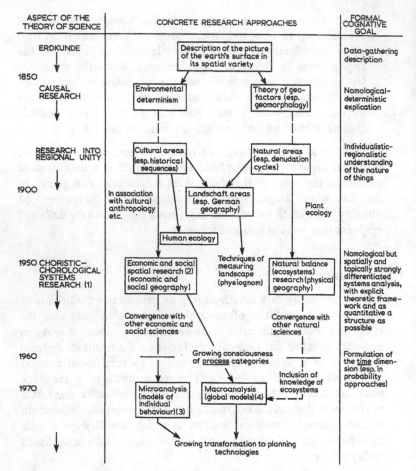

Figure 2.1 The geographical research approaches and their formal cognitive goals during the last one hundred years set against general aspects of the theory of science.

NOTES TO FIGURE 2.1

1. Complex theories, operationalised in the framework of systems analysis, for whose components differing locations are set out on the earth's surface (choristics: the description of distributions on the earth's surface) and for which it is possible to identify distance-dependent modifications mediated by

(Gehlen), which is necessary for the perception of 'normal' everyday questions of research. Hard and rigorous introduction of the fundamental concepts of a basic perspective and its component theories is, therefore, not simply authoritarian capriciousness in study, but the force of an acculturation process within research on whose progress depends the liberation of energies for the solution of individual scientific tasks. The advocates of a permanent and total critical attitude in learning neglect this situation in the course of a corresponding over-estimation of the individual researcher's capabilities.

3. See also, for example, Gould (1969); King (1969); Kohn (1970).

4. Harvey (1969); also Pred (1967/68); Olsson (1970 and 1971); or Rushton (1969). Compare Luhmann (1968) for the analogy of similar basic problems of so-called structural-functional analysis in sociology and geography over and above the current simple systems theory formalism.

5. That attitude of consciousness which sees every societal objection and contradiction as a harbinger of catastrophe, and which wavers between despair and euphoric hope in portraying the future, straining after a meaningfulness of social phenomena which will finally solve all antagonisms is, historically speaking, an inheritance of the Romantic Age.

6. It is probable that the social critic would reject this comparison on the grounds that the dialectic of the quickly changing situation does not allow an assembly-line system to be responsive.

---

movements (chorology: the analysis of functional linkages or movement resistances, for which direct spatial distance is at least an indicator).

2. Based on the new image of man, brought to maturity between the wars, whose constitutive force is no longer merely passive reaction on environment, but active *social interaction* (Weber).

3. Analysis of human decision processes and their determinants with special attention to the perception of space and environment. Component problems include: generation of group norms, interaction stereotypes, interaction in a situation of no knowledge or uncertainty.

4. Construction and verification of regional systems with elements of aggregated size (from human activities and their peripheral constraints) with identification of fixed or manipulatable parameters; that is, e.g. technological constants and/or planning flexibilities. Component problems: adequate aggregation procedures in spatial (e.g. region-building) and topical terms (e.g. groups or sectorial divisions).

In this way, however, normal research and teaching organised on a disciplinary basis would become completely impossible, insofar as here a realistic model of science is open to debate.

7. The circle which lies in this explanation (what is relevant is that which is relevant) can only be eliminated by rounding out the content of the dimensions in a manner of joint engagement.

References

ALBERT, H. (Ed.) (1964) *Theorie und Realitaet*; (Tuebingen).
ALBERT, H. (1968) *Traktat ueber kritische Vernunft*; (Tuebingen).
BARTELS, D. (1968) *Zur wissenschaftstheoretischen Grundlegung einer Geographie des Menschen*; (Wiesbaden).
BARTELS, D. (Ed.) (1970) *Wirtschafts- und Sozialgeographie*; (Koeln).
BERRY, B.J.L. (1964) Approaches to regional analysis; a synthesis; *Annals of the Association of American Geographers* 54, 2-11.
BERRY, B.J.L. (1968) Interdependence of spatial structure and spatial behaviour: a general field theory formulation; *Papers of the Regional Science Association*, 21, 205-227.
BOBEK, H. (1957) Gedanken ueber das logische System der Geographie; *Mitteilungen Osterreich. Geog. Gesellsch.*, 99.
BOBEK, H. (1962) Uber den Einbau der sozialgeographischen Betrachtungsweise in die Kulturgeographie; *Verhandlung. Deutscher Geographentag Koeln 1961*, (Wiesbaden).
BROOKFIELD, H.C. (1969) On the environment as perceived; *Progress in Geography*, 1, 51-80.
BUNGE, W. (1962) *Theoretical Geography*, (Lund).
BURTON, I. (1963) The quantitative revolution and theoretical geography; *Canadian Geographer*, 7, 151-62.
CHORLEY, R.J. and HAGGETT, P. (Eds.) (1967) *Models in Geography*; (Methuen, London).
CLAVAL, P. (1964) *Essai sur l'evolution de la geographie humaine*; (Paris).
CLAVAL, P. (1967) Geographie et profondeur sociale; *Annales Economies Societes Civilisations*, 7 p. 1005-46.
COLE, J.P. and KING, C.A.M. (1968) *Quantitative Geography*; (Arnold, London).
COX, K.R. and GOLLEDGE, R.G. (Eds.) (1969) *Behavioral Problems in Geography: a symposium*; (Evanston, Illinois).
CURRY, L. (1967) Quantitative geography 1967; *Canadian Geographer*, 11, 265-279.
DOWNS, R.M. (1970) Geographic space perception; past approaches and future prospects; *Progress in Geography*, 2, 65-108.

GERASIMOV, I.P. (1970) Future tasks of mankind and the responsibility of geosciences; *Geoforum*, No 1, 71-6.

GOLLEDGE, R.G. and AMADEO, D. (1968) On laws in geography; *Annals of the Association of American Geographers*, 58, 760-74.

GOULD, P.R. (1969) Methodological developments since the fifties; *Progress in Geography*. 1, 1-49.

HABERMAS, J. (1968) *Erkenntnis und Interesse*; (Frankfurt).

HAGGETT, P. (1965) *Locational Analysis in Human Geography*; (Arnold, London).

HARD, G. (1969) Die Diffusion der "Idee der Landschaft"; *Erdkunde*, 23 p. 249-64.

HARVEY, D. (1969) *Explanation in Geography*; (Arnold, London).

HARVEY, D. (1970) Social processes and spatial form: an analysis of the conceptual problems of urban planning; *Papers of the Regional Science Association*, 25 p. 47-69.

HOFMANN, W. (1968) *Universitaet, Ideologie, Gesellschaft: Beitraege zur Wissenschaftssoziologie*; (Frankfurt).

KING, L.J. (1969) The analysis of spatial form and its relation to geographic theory; *Annals of the Association of American Geographers*, 59, 573-95.

KOHN, C.F. (1970) The 1960's; a decade of progress in geographical research and development; *Annals of the Association of American Geographers*, 60, 211-19.

KUHN, T.S. (1962) *The Structure of Scientific Revolutions*; (Chicago).

KULS, W. (1970) Ueber einige Entwicklungstendenzen in der geographischen Wissenschaft der 2. Haelfte des 19. Jahrhunderts; *Mitteil. Geogr. Gesellsch. Muenchen*, 55 p. 11-30.

LUHMANN, N. (1968) *Zweckbegriff und Systemrationalitaet*; (Tuebingen).

LUKERMANN, F. (1965) The 'calcul des probabilities' and the école francaise de geographie; *Canadian Geographer*, 9, 128-137.

NEEF, E. (1967) *Die theoretischen Grundlagen der Landschaftslehre*; (Gotha-Leipzig).

OLSSON, G. (1969) Geography 1984; *Department of Geography, University of Bristol, Seminar Papers*, Series A, No. 7.

OLSSON, G. (1970) Explanation, prediction, and meaning variance: an assessment of distance interaction models; *Economic Geography*, 46, 233.

OLSSON, G. (1971) Corresponding rules and social engineering; *Economic Geography*; 47 p. 545-54.

PRED, A. (1967/69) *Behavior and Location: Parts I and II*; (Lund).

RUPPERT, K. and SCHAFFER, F. (1969) Zur Konzeption der Sozialgeographie; *Geographische Rundschau*, 21 p. 205-14.

RUSHTON, G. (1969) The scaling of locational preferences; *Annals of the Association of American Geographers*, 59, 201.

TOPITSCH, E. (Ed.) (1965) *Logik der Sozialwissenschaften;* (Koeln).

TOULMIN, S. (1961) *Foresight and understanding: an enquiry into the aims of Science;* (London).

UHLIG, H. (1970) Organisationsplan und System der Geographie; *Geoforum*, 1, 19-52.

WIRTH, E. (1969) Zum Problem einer allgemeinen Kulturgeographie; *Die Erde*, 100, 155-193.

WITTGENSTEIN, L. (1963A) *Philosophische Untersuchungen;* (Schriften, Frankfurt).

WITTGENSTEIN, L. (1963B) *Tractatus Logico-Philosophicus;* (Schriften, Frankfurt).

ACKNOWLEDGEMENT: The author wishes to express his thanks to the publishers for arranging for the careful first translation of this paper.

3 · Theory of geography

V.A. ANUCHIN

One can trace definite cycles in the history of world science. Periods when the general absorbs the particular are succeeded by those during which the particular destroys the general and a single science disintegrates into an endless number of branches. This later differentiation leads to great extensions of knowledge but also results in the loss of integrative over-views of science which show that the whole is greater than the sum of the parts. Contemporary geography is a victim of such a phase of differentiation. Evidence of this is that, although there have been successes in the study of separate components of the geographical (or Ландшафт*) sphere, it is evident that the science of geography has not been able to solve the present important integrated social problems. Chief among these is how to use both the total human environment and its separate territorial complexes.

Is there such a science as geography?

The disintegration of geography has reached the stage that even some geographers ask themselves "Isn't the old lady dead?". There are many publications which, from various standpoints, attempt to discredit geography as a science. Geography has been vulnerable to such criticisms because, firstly, as its function is largely one of synthesis, it has greatly suffered during a recent phase of differentiation and, secondly, because it occupies a place between the natural and the social sciences.

Such a suggested liquidation of geography, sometimes proclaimed by eminent geographers themselves, certainly does not encourage its development. Since the beginning of this century geography has been constantly on the defensive with respect to other sciences and sometimes treated merely as a secondary school subject. This attitude has persisted to the present day and many scientists (like V.V. Dokuchaev), who were substantially geographers, have often

* Landschaft (Author)

preferred to give their activities another name! There has even arisen a special substitute terminology involving "regional studies", "science of distribution of productive power", "science of zones and regions", and the like. When one recalls that for decades geography was not recognised as a scholarly discipline, one cannot perhaps be surprised that some scientists have unsuspectingly spent their whole careers studying geography — rather like the well-known character of Moliére who did not know that he had been speaking prose all his life!

Of what is geography comprised? Its identity lies mainly, but not exclusively, in its *theory and methodology* for specialisation in science is progressive only if studies in depth in one branch are theoretically connected with those in neighbouring branches. When geographers forget this they inevitably lean towards some adjacent scientific discipline, or, as Thor Heyerdahl remarked, "they plunge deeper and deeper into their own hole so that they cannot see each other any more, and the results are neatly stacked on top". Each separate discipline, of course, is unable instantaneously to reflect the subject of its study in its totality. The deeper the knowledge, the more the discipline becomes subdivided and variously defined in different stages of its development by its specialists. Such restricted studies in depth may even result in an impoverishment of scholarship, because in itself fragmented knowledge does not enrich science. The way out of the current empirical crisis lies in the strengthening of *abstract thinking* that is able to fuse the scattered information and conclusions of specialist investigations into an elegant logical system and make possible the acceptance of empirical results as the content of a legitimate discipline. In other words, only by abstract thinking is it possible to synthesize the results of specialist observations.

The methods of empirical investigation, while inevitably strengthening analytical ways of thinking, in no way assist such an integration. The analytical division of separate parts of the subject leads to the creation of a logical system of empiricism, which, although useful in analytical studies in that it secures knowledge in the framework of a differentiated analysis, does not lead to an integration of scholarship. This can only be achieved by theory which results in the 'organic' coordination of results of experimental analysis. A theory must be an abstract method differing from the mere registration of the results of direct observation and experiment, be able to give a fuller understanding of the science as a whole, and be able to supply scientists with the dialectical logic of their discipline. After this has been achieved differentiation will obviously continue to dominate, but it is not an absolute domination.

The tendency towards the integration of different disciplines is now gradually becoming the most characteristic feature of science, as the necessity for a synthesis of accumulated analytical empirical knowledge becomes more pressing. The logic of any science lies in the definition of its substance and of its methodological basis. Such a logic can only form a really sound theoretical conception when it is based on an adequate definition of the subject matter of the science, not only at present but in the foreseeable future. Sound theory must be predictive; not only relevant to the present levels of science but showing the ways for its further progress and making scientific prognosis possible.

Theory must thus go much further than experiment in providing a new, more profound, and substantially a multi-lateral generalization of knowledge expressing the active penetration of human thought into an objective material reality. Theory goes beyond direct observation in its aim of revealing the essential relationships which define the character of the object of study. Theory points the way to further studies, joining and cementing the efforts of specialist scientists, and is the means by which a science is able to overcome the dangers of complete dismemberment.

However, although a theory which was productive at one stage of the development of a science can become antiquated in the subsequent stages, the theories of different historical periods are the steps of the progress of knowledge which is achieved through a summation of relative truths. In opening new horizons of knowledge, theory must itself develop, or else it will become its own antithesis and will be transformed into dogma.

A preoccupation with model building sometimes results in identifying this *method of organizing knowledge* with theory, although a model is unable to disclose the theoretical substance of the object being studied. Model building only widens the possibilities of knowledge, supplementing direct observations and experiments by indirect ones. It permits one to identify the stage reached by existing ideas and to predict the disclosure of new phenomena; it creates a new hypothesis about the character of the subjects studied and highlights the defects of existing theories. But all this does not transform modelling into a special theory and it remains only a *method* which must be augmented by theory.

The subject matter of geography

The subject matter of geography within the geographical sphere of the earth appears as a synthesis of all near-surface spheres into one interacting system. These consist of the lithosphere, the hydrosphere (including the bottom of the seas and oceans), the atmosphere, the biosphere and the sociosphere (or neosphere). The latter, like the biosphere, unites people with their environmental complexes and is a mutually-reacting component of the geographical sphere. The geographical sphere is thus a complex of systems developing through the mutual influences of different kinds of phenomena (physical, chemical, biological and social), wherein the accelerating interactions of man and the rest of nature have led to its "humanising" and to its transformation into a medium for social development. One can divide landscape complexes into those which originated as the result of spatial variations which existed before man and those created by man, the latter based upon man's material additions and changes to the natural landscape.

Man's territorial influences have developed progressively and at first there were only small ecumenes* inhabited by human communities. These were later extended and at present it is difficult to find large land territories which are not being exploited in the interests of humanity. Moreover these unexploited territories have long interacted with the ecumenes in that, for example, uninhabited mountain tops influence the distribution of moisture over large areas and the prevalence of ice in the polar regions affects the world ocean level. The whole of the geographical sphere can thus be considered as the environment of human society, but at the same time its various

---

* In this instance the term ecumene is used to show the "hearth" character of the early geographical environments and their subsequent territorial extensions.

parts should be identified by the degree to which they are exploited in terms of human production processes. Although it is hardly possible to view the ocean bottoms as a present geographical environment, a time will come when even this part of the landscape sphere will be embraced by human activity and brought into economic production. The geographical environment is a narrower concept than that of the landscape sphere, since it includes only such parts of the latter that have become directly related to the life of human society, although, potentially, the whole landscape sphere should be viewed as the geographical environment.

The total geographical sphere of the Earth can therefore be subdivided in terms of the relative degrees of the influence of society:

1. Those parts which are outside social activity and the development of which depends on the operation of physical, chemical and biological laws.
2. Those parts of whose regional complexes experience indirect social influences and which are strongly controlled by physical, chemical and biological laws. These regions are those which have no stable population and the influence of man is almost exclusively external.
3. Those parts whose regional complexes experience both indirect and direct social influences, and which form part of the socio-geographical environment. The development of these is controlled by the interactions of physical, chemical, biological and social laws.
4. Those parts whose regional complexes have been directly involved in production. Here the operation of natural laws occur under strong social influences and only roughly define the limits of development. Here the interaction between different classes of social, biological, and physico-chemical laws is of paramount importance. These form the remainder of the socio-geographical environment.

It is thus becoming necessary to develop indicators of the measure of saturation of the geographical sphere resulting from the interaction of society and nature.

The relations between society and the environment presuppose mutual interactions between them. When speaking of contemporary humanity you certainly cannot call the development of its environment purely 'natural', for people are actively forming their environment. All living organisms create for themselves an environment of biosynosis, coexistence, seasonal rhythms, etc. and humans are not an exception. Human society creates an environment and in this process apart from production (which undoubtedly is the leading factor) a great part is played by life activities of people as biological individuals. The definition of "society" (until a new definition has been found) thus becomes rather specific in geography. It is important for the geographer to see society not only as a social phenomenon, but also as a biosocial one.* In studying social phenomena we must not limit ourselves to the operation of purely social laws or production relationships that attribute to society alone the internal causes of its own development. "For geography, besides all the social elements, people are important both as a definite volume of biomass in the geochemical balance of the planet and as biological consumers of enormous quantities of living and non-living material of the crust of the earth, and of the hydrosphere and atmosphere, and as suppliers of the carbon-monoxide and other rejects of the activities of its organism which are returned by them to nature. The biological substance of humanity perceptibly influences the fate of its social development. The requirements of food, warmth, multiplication and support and continuation of life continue to be the foundation of the most basic social processes, serving as the prime causes of labour, defining its necessity, and leading to the gathering of people into collectives" (Efremov 1968, 95). The landscape sphere must be studied as a complicated combination of developing systems, each containing certain of the attributes of the preceding less-advanced systems. "The fact that each subsequent component of the geographical envelope represents

---

* Society, as a self-developing material system, forms the subject matter of historical materialism. Society is connected with other systems, but the causes of its development have an internal character. It differs from other systems in possessing consciousness, from which arises the first problem of the interrelationship of social existence and of social consciousness. Society includes everything which is drawn into production, i.e. people possessing definite knowledge and qualifications, the means of production, the conditions of work and life, the means of labour, and the products of labour.

a product of all already existing ones is a very striking proof of the theories regarding the unity of the geographical envelope. Having been created, it always functions, develops and gives birth to new ones as a unity and not as a sum of components" (Blauberg 1960, 237).

The interrelationships between human society and its environment are still obscure and it is often wrongly held that society and environment must not only be opposed to each other but must be studied independently. Man is a part of nature and is limited by it so that humanity is not only a social, but also a biosocial concept. Man will always remain bound to the rest of nature by unbreakable ties. The qualitative differences between society and the rest of nature permit one to study their interrelationship either as interrelationships between subject (society) and object (nature), or as interrelationships between two parts of a general whole, presupposing, however, that differences exist between society and the rest of nature. This latter view forces one to consider the interrelationships which exist between social and natural laws. Such a specific method of approach to the study of the substantially material aspects of society divides geography from the other social disciplines that study society as an isolated whole and endeavour to establish laws inherent in it. If our point of departure is the understanding of the unity of material world, then it is impossible not to recognize a unity between natural science and social science although not denying important differences in principle between them. Social studies must rest on natural foundations which may have been altered or even created by man.

At present an ever-increasing number of scientists are coming to understand the unity in principle of physical and economic geography. K.K. Markov (1965, 59-60), referring to the unity of geography wrote: "The first step must consist of recognition of very close ideological ties based on Marxist theory between the two basic branches of the science of geography — physical and economic geography. Enormous efforts have been directed to a negative purpose — to provide the weakness of these ties and to reject the ideal of indivisibility of geography. . . . A negative decision would represent a suicidal act and would lead to the decay of geography as a science and as a unit in teaching disciplines." The study of "dehumanised" nature is thus decreasing in practical importance as the history of nature and the history of human society become more and more interdependent and join into one process. In a century of great technical revolution, and particularly when studying the physico-geographical conditions and resources of Soviet Russia, it becomes impossible to isolate oneself from the laws and demands of social development. In order to study the nature of agricultural regions, for instance, one must consider the history of cultivation in a region, its present character, the foundations of agricultural science, as well as the economic aims for the future. When studying industrial regions the physical geographer must understand the mineral as well as the surface resources, replacing the bio-geographical zonation of such regions by geochemical zones connected with sources of useful minerals. In extreme instances local water resources, soil, vegetation, and animal life can be completely destroyed by industry, roads and towns; even the relief of the surface can be changed. (Kolosovsky 1955, 142).

The study of natural complexes is of practical value only when its results are considered as a basis for economic activities, and this is possible only if a study of the natural and social complexes of the geographical environment is undertaken simultaneously. With the growth of man's technical skills, the geographical environment acquires a more and more anthropogenic character, but these changes do not annul the action of the laws of nature. They merely create new links between society and nature thereby increasing the importance of the general geographical approach. The unity of geographical science is achieved by means of the unity of the subject studied — the geographical environment.

The socio-geographical environment

The definition of that part of the geographical sphere which has become the milieu of social development, as well as of the problems connected with the creation of a new theoretical conception of geographical science, prompts the following observations:

1. The socio-geographical environment is that part of the geo-graphical sphere in which the development of human society takes place in direct interrelationship with the natural environ-

ment. It is not the limitless world of the whole of external nature but a specific part of it.

2. The most characteristic feature of the socio-geographical environment is that it has been modified by human activity directed to definite purposes, such that it is being to some extent saturated by the results of this activity at a rate that is immeasurably greater than those affecting the rest of nature.

3. The elements of terrestrial nature involved in the process of social production are transformed into the productive forces of society, into its working tools, into the results of work and into social elements of the socio-geographical environment. These productive forces form part of the socio-geographical environment but they do not compose it completely. The process of involving new parts of nature with economic production is a social process, which must employ the laws of nature in a way which does not contrast social and natural laws. The development of the socio-geographical environment thus take place not only as a result of external natural influences (e.g. the receipt of solar energy) and of internal natural processes, but also as the indirect results of social phenomena as the processes of production.

4. The socio-geographical environment is that within which are accumulated the material means of social development, but it is at the same time the result of the social action on nature. Only in this sense can one speak of the unity between society and socio-geographical environment which brings about a "humanised nature" (Morozov 1963). When the exploitation of cosmic distances becomes more possible one of the chief problems will be the creation of socio-geographical environments on other planets. Ultimately these will be created out of the natural materials existing there, but at first materials will be exported there from Earth. Therefore the process of changes in the socio-geographical environment achieved by society grows from epoch to epoch in accordance with social requirements and determined by methods of production. The socio-geographical environment is an historical feature.

5. The complicated character of the socio-geographical environment also complicates its study. It cannot be undertaken from the point of view of a single geography which does not see qualitative

peculiarities of society, or which does not take into consideration
the influences of methods of production on the whole character
of the social pressures on nature. Nor can it be studied by looking
upon society as a result of outside material relationships, rather
than as a totality of production relationships.

6. The socio-geographical environment is a complex combination of
natural and social conditions. The mutual influences of society
and nature do not exist in reality, being abstract phenomena
existing in our consciousness and being represented in real life
only by their material expressions. Because one type of economic
activity commonly results in the establishment of others in a
complex way, one cannot view the socio-geographical environment as the cause of social development, as was thought by
deterministic geographers, but *vice versa.*

The methodology of geography

Geography is concerned with the existing material system forming
the geographical sphere of the Earth as an environment for the
actual or potential development of human society, together with the
material aspects of social development expressed in its regional
complexes within the geographical environment. It follows from this
that a spatial approach is the methodological basis of any geographical study. This regional approach inevitably involves the
historical method, but this methodological unity of the geographical
and historical sciences does not deny important differences in
principle between geography and history. The link between
geography and history is not entirely a methodological one.
Geographers, like historians, reproduce in their consciousness the
processes of development of a complicated object with the aid of an
aspect of the dialetical method. The geographical method concerns
itself with the complicated developing systems of links which have
been, or are being, formed differently in different regions. Logic in
historical science is concerned with the reproduction in the mind of
processes of development of society and nature, expressed in time;
logic in geographical science with the processes of development of
the geographical environment expressed spatially over the earth's
surface. Both the development in time and the development in space

are constantly linked together, although they are far from being always synchronised, and one must remember that the geographical logic embraces both aspects of the existence of matter, time and space. When studying the regularities of development and form, in which it is expressed in time, history cannot divorce itself from geographical conditions, as specifying the location where this development took place. It follows from this that historical science also employs the geographical method. Without the geographical method it becomes impossible for historians to ascertain the many local features in historical development, and their work will suffer from abstractness in which many important specific historical processes will be overlooked. However, although the ability of an individual discipline to employ many methods is strong evidence of the unity of the processes of knowledge, each discipline always has its own *fundamental method,* its methodological basis.

The study of the methodological specifics of geography is thus no less important than the definition of its subject and is at present particularly important in view of the penetration into geography of new techniques, especially of mathematics. During this penetration it is important to keep in mind the *methodological foundation* of our science, even though improved means of perception may assist in the solution of its new specific problems. If the technique is unrelated to geographical methodology it will be a foreign body, encumbering with unnecessary abstractions and symbols the process of geographical perception.

The penetration of mathematics into geography cannot lead to a creation of a new branch of the discipline because the introduction of a new *method* does not automatically result in the creation of a new *subject* of research. However, far from being used only for solving particular problems, mathematics introduces possibilities for new and deeper comprehension of the problems under investigation and strengthens the *theoretical aspect* of geography. Better possibilities for generalization are presenting themselves, as well as of clarifying the common features of phenomena with different detailed characteristics, while at the same time revealing more precisely their individual features. Mathematics is thus a good medicine against an illness known as "the inflation of specialisation."□One of the reasons often put forward against the possibility of wide-ranging studies of general geographical synthesis, revealing the general systematics of development and distribution of complexes of the socio-geographical environment particularly those relating to the productive forces of society, is the growing difficulty of absorbing the enormous amount of factual material. One hears that "Geography is at best a descriptive science which is moribund because of its fragmentation and because it is not able to generalise the regularities of space, or even to give precise characteristics to one or other territorial complexes." I do not agree with these kind of arguments. Even using its traditional methodological weapons, geography is able to do far more than merely describe things. But undoubtedly the penetration of mathematics strengthens the possibilities of development of wide and, at the same time, precise studies. Mathematics is being transformed into the "esperanto" of scientists, the formalisation of problems increases the mutual understanding amongst geographers of quite different branches, and this secures a better knowledge of the whole discipline. Geographico-mathematical methods open new possibilities for general geographical studies. In particular spatial modelling, in combination with cartography, has the widest application, particularly in investigating the complex exploitation of productive resources.

The problem of mathematization is one shared by many contemporary sciences which are being submerged under an enormous quantity of empirical data. A satisfactory solution of this problem can only be reached by grafting these techniques onto the methodological basis of geography. However, the use of mathematics in geography will always have definite limitations because mathematical formulation is unable to fully treat the multivariate environment with which geography is concerned, embodying all the unrepeatable combinations of elements within its territorial complexes. In perceiving the socio-geographical environment one can observe two basic tendencies. In one case the general forms with their logical analysis are distinguished; in the other, basic attention is given to the study of local phenomena which cannot be integrated into the general forms and which impede their logical analysis. In geography this second tendency involves a constant probing of geographical specifics at different spatial scales, and, as such specifics are individual and non-repeatable, their mathematical generalization is not possible. More than this, the analysis of geographical phenomena can even be obscured by mathematical techniques which may hide their full clarity, completeness and individual non-repeatable character. N.N. Baransky (1956) was right when he showed a timely concern for the art of geographical description, for description is most necessary in all stages of the development of geography. The methods used in geography have not been exhausted and must be applied and developed with the aid of qualitative and mathematical techniques. It would be a great error to think that only qualitative methods, or the use of mathematics, define the ultimate scientific level of a particular geographical work. The change to quantitative approaches to scholarship does not obviate (but on the contrary presupposes) the application of qualitative methods, and already we can see a swing back from quantitative methods again to qualitative, but on a new higher level. Such is the dialectic of perception, and it is quite wrong to set the two approaches in direct opposition to each other. All methods are good which permit a broader mental perception of the unknown world.

Geographical phenomena comprise more or less simple forms commonly amenable to graphical treatment, allowing quantitative analysis which permits the mathematical method to be applicable in establishing the structural unity inherent in them. During any regional investigation not only the systematic but also the general characteristics of the geographical sphere of the Earth and of its individual regional complexes become clear, and it follows from this that, not only in selecting the primary material but also in its subsequent handling, the mathematical method can achieve much. In particular, the study of the dynamics and relationships of individual phenomena can be pursued with great success with the aid of mathematics. Mathematics can, and should, be used in all cases when the questions facing one are sufficiently complex to demand the creation of their own mathematical symbolism and specialized sequence of solutions (algorithms). It is now becoming impossible to analyse logically the accumulated mass of factual material without algorithms. First of all the unknowns are defined and the problem is posed; then its boundary conditions are formulated, the primary information is studied, and the criteria are chosen; finally, the sequence of the rules of solution are established and programs compiled. Usually dozens of programs are prepared in advance out of which a suitable one is chosen for the existing problem. Any solution which is obtained as the result of applying mathematics in this way will not fully or correctly represent reality and, when the number of constraints taken into consideration is small, it can be absurd. Therefore any solution cannot be further applied without a preliminary analysis which involves its testing on different levels and variations, and consideration of the qualitative peculiarities of the factors and the manner of their operation. Only such combinations of solutions through geographico-mathematical analysis leads to the satisfactory solution of problems in the field of regional studies. Mathematization supposes a simplification producing models of reality, the scientific and practical importance of which becomes greater with the increasing number of essential factors and conditions which are taken into account. Here we meet with the contradictions of any modelling operations. Being a simplification itself it must avoid oversimplification! Therefore it is very important to define the permitted limit of simplification for each problem. In the sphere of regional studies this is established on the basis of study of reality through time by qualitative, and not by quantitative, methods.

When deciding any problems of regional character, particularly in the distribution of productive power along with a quantitative evaluation of different factors, a great role is reserved for qualitative evaluation, which as a general rule can be expressed only poorly or not at all by quantitative indicators. (This is applicable particularly to regionally non-repeatable geographical phenomena.) Qualitative evaluations are thus extremely important even when defining the effectiveness of capital investment into the development of some territorial complex; not to mention the problem of correct exploitation of labour resources. The valuation of invisible benefits which are difficult to estimate is often no less important than the valuation of tangible benefits. For instance, when analyzing the effectiveness of capital investments in the eastern regions of the USSR it would be wrong to arrive at conclusions only on the basis of comparison with the present effectiveness in the more-developed western regions. If, however, one takes into consideration the advantages for the entire present and future domestic economy of the country, then the siting of many industrial projects in the eastern regions will be fully justified, despite the fact that quantitative calculations may indicate the opposite. The problem of the siting of national productive units commonly requires quite different decisions than those arrived at purely by quantitative analysis. The application of mathematics to the analysis of specific problems can be successful when the application of quantitative criteria is obligatory. However, it is more important to state the problem correctly and fully, than in a simpler manner which is amenable to mathematical treatment.

Regional modelling

Regional modelling as a method of investigation is certainly not new, and has been applied in geography more or less from its inception, often by means of the mapping of information. But every regional modelling presupposes a preliminary qualitative study in order to select the most important data and to formalize them correctly. Before modelling can be carried out qualitative simplification must take place in which the general dominates over the particular and the knowledge of the particular is gained through the general.

Geographers have to deal with complicated dynamic systems, between the reality of which and their theoretical models it is impossible to secure complete similarity in time and space. The synthesis and analysis of such systems requires the creation of a special information theory which up to the present has been far from adequately worked out. There is a need for criteria of similarity of purpose, of information structures and streams, and, finally, of the effects of information reorganisation at the centres of management. When modelling dynamic systems, where usually complete similarity is unattainable, one must take into consideration many factors which cannot be accurately specified from the mathematical point of view, but if these are ignored in favour of complete mathematical rigour then most geographical problems will become either insoluble or their solution will be unsatisfactory. The solution to any contemporary geographical problem is to a considerable degree based on the use of a whole system of models (graphical schemes, maps, geographical matrices, etc). In the first instance these are applied to the solution of questions connected with the problem of regional organisation of production, where synthetic models of regional systems of productive forces are in especially wide use. It is theoretically important to emphasise that any regional system is only relatively isolated, for any system is a subsystem of another more general system.

The domestic economy of a country is a system with a huge number of phase-space variables, consisting of many second- and third-order systems. The organs of administration (ministries, boards, etc) in no way influence all the phase-space variables, but only a certain proportion of them. At the same time the higher the level of the organ of administration the more variables are drawn into its orbit. This creates the problem of optimisation, the solution of which is required at every stage of the development of productive forces, and with every new stage it becomes more and more complicated. Therefore in contemporary geography the optimum modelling of the geographical environment acquires actual importance in that it assists in the definition of the optimum direction of development over time. Such a model makes possible the establishment of the optimal plan of development of the structure of the geo-environment, including the choice of time for the introduction for practical exploitation of natural resources, specifying at the same time controls over their usage in time as well as in quantity, and establishing optimal levels and procedures of production. This is of particular importance in forecasting the further development of regional productive complexes.

The study of the behaviour of a system is only possible if it is abstracted from the numerous individual features of a given real system and is based on some generalized concept, the fundamental characteristics of which are typical of real systems as a whole. It is important that this generalised system should conserve within itself the basic characteristics of actual systems which form the bases of a scientifically-justified abstraction which allow theoretical model systems to be related to actual complex systems. The elucidation of such generalised large and complex systems is of importance not only for clarifying the basic specific qualities of each type of system but also for securing an optimal organisation for their rational management. These basic theoretical aspects of modelling large and complex systems have so far been very little used in geographical studies, which have relied on other methods of regional classification. In the very near future, however, it can be expected that general geographical studies will be successful in applying modelling to regional planning studies as a prelude to the modelling of larger :geographical systems; i.e. for large natural and economic regions. When attacking the problems of regional planning it is particularly important to establish the optimal variations in the utilization of existing resources, the spatial distribution of productive and infrastructural objects, and the peculiarities of the labour supply. The variation sought should ensure the minimal expenditure on construction and exploitation, which usually means observing the following conditions:

(a) Ensuring a definite amount of production from using an established amount and mixture of resources.
(b) The volume of the attained production must correspond to established demand, taking into account imports and internal requirements.
(c) The territory to be built over, as well as the water resources, must accord with the established constraints, and the amount of harmful sewers and of gas outlets must not exceed the established maximum.
(d) The numbers of the active population must not exceed the established maximum. The living quarters must be separated from industrial areas within time limits established for travelling to and from work, and at the same time they must be outside of the boundaries of the harmful industrial influences.

Even in such a very generalised form the application of these criteria permits calculations to be made regarding the establishment of the locations sought for industrial undertakings, the scale of their development, the intensity of their intercommunications, the volume of their freight movements, the distribution of local resources between the undertakings, etc.

Fruitful applications of mathematics is taking place in those branches of geography which show the greatest need for the establishment of general relationships between the elements under study. This applies at present particularly to branches of physical and economic geography. The extension of these techniques to the social branches of geography is more speculative as qualitative considerations are more important here. One must not forget that qualitative differences between regional complexes are often connected with differences in the social make-up of life, which hinders the development of generalisation relating to regional complexes developed under the agency of different methods of production.

It should also be borne in mind that the mode of production plays an active role in the relationships of society with the rest of nature.* Thus modelling of territorial complexes will bring wrong results if the influence of relations of production on the production process is omitted.

Theory and practice

The absence of a sound theoretical basis in contemporary geography is one of the chief reasons of the inability of geographers to answer questions that arise in practice. The narrow development in the specialisms of geography assists penetrating analyses of separate elements of the geoenvironment and the accumulation of factual material, but few general geographical conclusions result from these analyses. It is often said that theory is not at all necessary in the solution of practical problems and that geographical studies should be concerned with contemporary practical needs. Science is a special activity but this does not mean that theoretical studies can have no direct practical importance, however long delayed this may be. To serve the day-to-day practical needs of society does not mean that there should be no theoretical work; on the contrary, it is the concentration on theory that safeguards the practical importance of science. To serve these needs scientific theory must go in advance of practice, predicting and guiding its development and, by so doing,

* Among factors that determine the development of the geographical sphere of principal importance are: *the mode of production* (determines the *cause* of development), nature of the productive forces, peculiarities of territorial combinations of the productive forces and effects of the laws of nature, peculiarities of their influence (determine the *conditions* of development of the geographical sphere).

establishing society's needs. N.N. Baransky (1960, 16) has emphasized that a narrow practicality impedes the development of science and that theory has given considerably more to practice than the most 'practical' work has. As evidence of this it is enough to compare the work of Faraday and Maxwell, on the one hand, with that of Edison, on the other. The total pursuit of burning contemporary questions of practicality results in an insufficient appreciation of theory, for running after today leads you into yesterday. Theory is not a mere reflection of practice, although sound practice results from sound theory.

The present day faces geography with very complicated problems, the solution of which is hampered by the inadequacies of our science primarily in the field of theory. Geographers will be unable to satisfactorily answer contemporary questions as long as they approach them from the standpoint of a differentiated, over-specialized geography. In the meantime the increase in population, in the need for food and an ever sharper increase in the need for every kind of raw material are part of a single process. The unity of society with its surrounding nature is realised in the regional complexes of the geoenvironment and the interdependence of society and environment forces us towards a strict evaluation of the relationships between them. A time is coming when any economic and technical calculations of the effectiveness of production will be valueless if they have not taken into consideration the unifying variables of the geoenvironment. Humanity has already approached a critical level beyond which the ignoring of the operation of the geoenvironment and of geographical science will lead to the ruin of the original basis of civilization and to the complete devaluation of all the economic advantages of the existing production processes. It becomes self-evident that production can only fully develop when it is based on geographical forecasting. Man's means of influencing nature have so increased that their application cannot continue without a study of their possible consequences. Geography, however, has not shown itself ready to solve this problem. The existing main branches (geomorphology, hydrology, demography, economic geography, etc), in spite of their usefulness, are completely unsatisfactory when one is concerned with questions connected with the evolution of the variables involved in regional complexes of the geoenvironment which control the possibilities of the development of production. For this purpose one must have synthetic general geographical studies the results of which would provide practical forecasts of the consequences of interference with natural processes, which inevitably are taking place. It is necessary to have a science concerned with the utilization of nature, a science which will connect natural science with the group of social sciences – what is needed is a geography without adjectives!

Economic development must be subordinated to the broader considerations of geographical studies. It must be modified, and in individual cases arrested, if it will lead to the destruction of the basis of life in a given region. In the past economic efficiency has not taken into consideration the protection of the geographical environment and is thus becoming its own antithesis by leading to the destruction of the fundamental riches of humanity. This applies primarily to air, water and soil. Oil, coal, metal ores and timber, although now indispensable, will in the future be substituted by other materials; but, without oxygen and water, life in its present form is impossible. Air and water, together with the land surface, are absolutely necessary as conditions and resources of life and for the further development of humanity. It might seem that this is an elemental truth obvious to all, but it is precisely these resources that have been subjected, and continue to be subjected, to a most irrational exploitation and pollution in just about every country in the world. There have recently been many publications on the theme of pollution of the environment, mostly of a non-geographical origin, and geographers are increasingly finding themselves outside one of the basic contemporary problems of geography! Geography finds itself unable to solve the problems of synthesis required to establish principles for the utilization of the geographical environment. Differentiation has thus brought about a loss of understanding of the substance of geography. We now see that contemporary geographical problems are beginning to be solved by the representatives of other sciences. This not only confirms the absence of impenetrable barriers between the sciences, but gives evidence of the methodological bankruptcy of those scientists who, although belonging to a 'geography department', have in fact been engaged in anything but geography!

References

ANUCHIN, V.A. (1960) *Theoretical Problems of Geography.*
ANUCHIN, V.A. (1972) *Theoretical Fundamentals of Geography.*
BARANSKY, N.N. (1956) *More Care of Art in Geographical Descriptions.*
BARANSKY, N.N. (1960) *Teaching Method for Economic Geography.*
BLAUBERG, I.V. (1960) *Philosophical Questions of Nature Study.*
EFREMOV, U.K. (1968) *Landscape Sphere and Geographical Environment.*
KOLOSOVSKY, N.N. (1955) *Scientific Problems of Geography*, 'Questions of Geography'.
MARKOV, K.K. (1965) *Space and Time in Geography.*
MARKOV, K.K. (1965) *Geographical Science and Advanced Geographical Studies in Universities.*
MOROZOV, N.V. (1963) *Society and Nature as Parts of a Whole.*
NEKRASSOV, N.N. (1972) *Economic Policy of the CPSU and the Location of the Productive Forces.*

ACKNOWLEDGEMENT: Thanks are due to the Cambridge Philosophical Society for arranging for the careful first translation of this paper. The final version was produced by the editor and publishers in collaboration with the author.

PART II
Spatial

4 · The domain of human geography

TORSTEN HÄGERSTRAND

The need for some redefinitions

'Human' in human geography implies a subject matter; 'Geography' implies a way of viewing this subject matter. The trouble with advances in human geography is for the time being perhaps not so much that we have failed to understand man's needs and wants as that our geography is too incomplete to be able to catch the conditions which circumscribe man's actions. If this is accepted follows that our present task is not to borrow scattered ideas from other fields and try to apply them within our present frame of geographic thinking. What we should do is rather to look for possibilities to develop the frame as such in order to make it more productive. To do this is not to try to create a new discipline. I believe that we have to admit a distinction between what should be accepted as geographical in kind and what is presently geographical in format. The latter, representing the conventions of the profession, should not be permitted to define the former. Advance is very much dependent on our ability to free ourselves from existing constraints when looking for the seeds of a theory which can subsequently grow by collective work.

Much of my critical attitude to the way in which human geographers have defined their task comes from experience in trying to make geographical thinking useful to policy-makers and planners in issues such as regional and settlement policy at the national level, national land-use control and future-oriented research. The two first questions are clearly geographical in character. Regional and settlement policy as a concept is an abbreviation covering everything involved in the formulation and implementation of a strategy for the distribution of population and their constructions and activities over the territory. The land-use budgeting is the projection of regional and settlement policy sieved through concerns for natural resources, recreation and the protection of the environment. General forward-looking studies represent, at least in their most long-range form, discussions of alternative ways of patterning social, political, technical and economic matters. Futurologists do not usually display much interest in using geographical perspectives in their reasoning, but it is quite obvious that 'scenarios' of future states of the world are fruitless unless their spatial dimensions are also considered.

When trying to apply geography one becomes very conscious of its present limitations and finds that there is a rather surprising degree of arbitrariness in its conceptual structure. Policy-making and planning cannot simply rely on a knowledge of the present state of affairs and its historical background, something which geography is good at describing. Nor is the mere extrapolation of trends, which we have sometimes tried to produce, an acceptable base for decisions on the societal level. The practical problem for the policy-maker is in most cases to change certain crucial relationships within an area or to add or subtract complete components in order to arrive at some wanted level of performance. Our arsenal of tools in human geography is indeed very modest as an aid in this sort of activity. We have not even given much thought to the evaluation of the performance of socio-economic systems as a function of their spatial arrangements. However, the geographical approach does contain a solid basis for developing instruments for just these sorts of tasks. If we tried to define the field of geography somewhat differently than we do at present it would in my view have a better potential to develop into a powerful analytical and creative aid in policy matters.

By giving weight to applications, I do not want to imply that we should let work in the discipline be governed by narrow practical demands. However, even those who are not themselves interested in application should appreciate that 'clinical' work has a lot to offer in terms of insights into processes at work and is one source among others, and perhaps one of the most productive, for suggesting theoretical structures. It is likewise important to stress that the scholarly elaboration of the findings must be given proper time and resources to be carried out quite separately from applications. Such basic research needs to go its own way even if usefulness is a final aim.

Let me first point out some of the boundaries pertaining to the conventional format which seem to be remarkably arbitrary, and after that proceed to the presentation of a few ideas about what could be looked for beyond them.

All geographers have in common an attraction for the out-door environment. "In size", say Abler et al. (1971) "the geographical scale of analysis and observation is bounded on the lower end by the architectural region – the area an architect usually considers when he designs a building – and on the upper end by the size of the earth." This is certainly a true description as far as most of the traditional human geography is concerned, but I do not believe that this is due to a conscious selection of scale interval. Rather, we passively accept the consequences of our long-standing outdoor interests. Contemporary armchair geographers are heirs to generations of explorers who viewed the world from saddles and sedan-chairs and of surveyors who depicted it on plane tables. We have learned to share their scale perspective and to favor their techniques of recording.

I am inclined to believe that the tradition coming down to us from exploration and surveying tends to cut off both the human and the geographical content of our investigations in ways which prevent the field from developing in a truly fundamental direction.

First of all let us consider our dominating instrument of recording. We require that observations be rendered graphically on maps. This habit is perhaps more risky today than it used to be because we have started to take the map much more seriously. Earlier the map used to be a means of general orientation and an adjunct to verbal discourse. Our private field experience helped us to extract much more synthetic information intuitively than was directly depicted sign by sign. Now the signs are the direct basis for serious and precise measurement. To no small degree the recent quantitative analysis in geography represents a study in depth of the patterns of points, lines, areas and surfaces depicted on maps of some sort or defined by co-ordinates in a two- or three-dimensional space.

I have no quarrel with this development. It is a great advance. But the real significance of it lies perhaps after all in the strengthening of our feelings regarding the value of a geometrical outlook. The danger

lies in its influence on how we come to view the relation between
signs and real-world phenomena. It could be that the map, by its
limitations, is selecting for us certain classes of spatial phenomena
and making us forget others.

Difficult problems of definition are embedded here with which I
am unable to deal rigorously. Let me at least try to direct the
readers' thoughts in a direction which I intuitively feel to be crucial.
I wonder if it is not true to say that we have been so exclusively
interested in the distributional arrangements of things and quantities
in a relative locational sense that we have tended to overlook the
space-consuming properties of phenomena and the consequences for
their ordering which these properties imply. The frequently-quoted
definition of human geography as a 'discipline in distance' – which
in its way is a good one – gives no hint of a concern for spatial
competition, for the 'pecking-order' between structures seeking
spatial accommodation. Even the classical preoccupation with *site* (in
terms of soils for crops and defensible hills for towns) which is
supposed to mean the opposite of *situation*, entails hardly more than
an aspect of relative location, although vertical instead of horizontal.
This lack of well-defined concepts and even ordinary imprecise
words dealing with interlocking, elbowing and predation as a process
in space is diagnostic of our traditional lack of concern for such
processes. The notion of space as made up of distances has
overtaken the notion of space as a provider of room. I do not mean
to say that the notions are unrelated but I see them as distinctly
different. The map is a poor instrument for depicting packing
problems except in the simpler cases (e.g. land-use). The map has to
be supplemented with some more abstract spatial notation before we
can develop an understanding of processes of this kind.

Given the recently-developed geometrical outlook it is rather
surprising that problems related to the 'grain-structure' of the
world have received relatively so little attention. There exist
scattered studies but they are few, and no systematic attempt seems
to have been made to attack the conceptual problem. Bunge (1969)
tells us that in order to be good optimizers we should "place
interacting objects of specified dimensions as near to each other as
possible". This statement is quite intriguing, not the least because of
some terms which have been left without comments. To say 'as near

as possible' implies that objects, whatever they are, are somehow prevented from coinciding. In other words, that objects, at least in some relations, compete for space. As soon as one object has found a location, the space it occupies is not available for a host of other 'weaker' objects and the probability field of their location has changed. Of course, objects are sometimes more and sometimes less closed and elastic relative to each other, and this makes the packing process very complex, in that the objects distort and penetrate each other in various ways. Whatever the complexities are, however, packing adds a meaning to interaction which is different from what we usually seem to have in mind, and certainly from what Bunge refers to in the quoted sentence when he talks about 'interacting objects'. Many objects in the real world interact just because they have come to be located adjacent to each other or one inside the other and for no other reason. This very kind of interaction can have great significance both for the structure of an area and for the sequence of events occurring there.

When members of other disciplines, or laymen for that matter, talk about the 'geography' of some area, they probably more often think of the mass of unassorted phenomena to be found there than of the orderly arrangements of selected sets which geographers recently have done so much to try to discern. This latter activity has been extremely productive but perhaps we ought to widen the repertoire to include the study of principles at work when unlike and *per se* unrelated things come together in a mix. I do not think we will have really covered fully our specific spatial point of view until we try to grapple with the competitive and cooperative processes at work in such a mix.

Leslie King (1969), in his short discussion of the packing problem, refers to isolated studies of the formation and survival of lunar craters, a choice of example which testifies the lack of concern among physical geographers. In human geography for example Hudson (1969) presents a longer discussion of space-filling as competition for land in his work on rural settlement patterns. However, both the case of lunar craters and of farms are concerned with how equals behave vis-à-vis each other. This, of course, is part of the story, but the real test lies in how we are able to handle the general case of un-equals.

It is easy to find examples from maps where natural and man-made features cannot co-exist in the same space without mutual distortion. A case in point is when road-systems are superimposed upon river-systems. They are both transportation networks with several traits in common. Their different nature as space-consuming objects must nevertheless be taken into account as soon as they are made to work within the same piece of space. Even today's road-building is to a considerable degree affected by hydrographic constraints and roads in their turn change at least the finer details of the water-flow. Who is looking into the principles of distortion when networks of different nature compete?

The picture gets vastly more complicated as we proceed to living organisms. Any meadow or piece of woodland stands out as an everchanging solution of the space-packing problem. Every plant, big or small, needs its minimum three-dimensional space above and below ground in order to survive. The individual plant cannot move from its base-point but, by its spacing-filling capacity up to a certain radius, it will compete or cooperate with the set of neighbours which have happened to take root around it. Even the supply of light is often a function of the manner in which plants mutually distribute shadows as the day goes on. Few, if any, geographers have tried to apply their tools to this microcosm. We have rather just borrowed a number of quantitative techniques from ecologists for much simpler tasks. Still here is a kind of universe of great interest also for the human geographer because the significance of space-packing processes is exemplified so clearly. There are probably certain similarities between ecological competition in biology and packing situations in urban areas, to mention a case from the 'normal' scale interval of the human geographer.

The neglect of space-packing processes could hardly have continued for so long time if we had not on the whole crossed out from our universe the microspaces conventionally belonging to the engineering and architectural scale. Because they are visible in the landscape we have not hesitated to analyse the function of for example equipments and fences in farming operations. But we have left unnoticed – except perhaps in terms of statistical indexing – the accommodation of machines and the flows of material, people, information and waste behind the urban walls, although these processes are of utmost importance for human conditions and probably mean a lot also for large-scale locational patterns because of constraints on arrangements imposed in micro-space. It is clear that the packing problem is impossible to abstract from in spaces below the size of the building. We have avoided these spaces and thereby lost the opportunity to gain insights which may have much validity also within the conventional scale-interval.

There is also a further geographical lesson to be learned from plant communities as well as from the urban-industrial micro-environment: one does not achieve any depths of understanding unless what happens over time is also considered. Thus it should be noted, with the subsequent discussion in view, that what I have called 'grain-structure' is not only observable in space but also over time. Organisms, machines and buildings form populations in which generations follow each other as parcels in time. Territories of all sizes are frequently bounded not only in space but also in time.

This observation introduces us to another feature of the traditional geographical format, namely the very strong emphasis laid upon the thin spatial cross-sectional view of the flow of terrestrial events. Of course, various efforts have been made to include a time-perspective of some sort, and historical geography in its most ambitious form tries to reconstruct the 'geography' of selected dates in the past. Change is then visualized as on a strip of film where the sequence gives some idea of long-term trends. Students of process often proceed in the same fashion, only that the time intervals become much closer. It is common to map some sort of index of change, calculated between two points in time. Models for making spatial extrapolation into future time have been developed but, on the whole, we have not yet succeeded in handling events as located and connected within a compact space-time block. Time is handled as a discrete entity. Again, to get out from this trap, a new sort of notation seems to be called for. Apart from the conceptual difficulties in handling the space-time problem it may be that we have become inhibited from trying to do so by the constantly-repeated assertion that geography is the science of flat spatial relationships as depicted on maps.

The second critical consideration deals with how man has come to be introduced into the geographic picture of an area. The tradition we have been left with is very indeterminate and very unsatisfactory in that respect.

It lies in the nature of his symbolic language that the surveyor cannot include mobile man among the more stationary objects of a landscape. Nor does it seem to have much informative value to try to locate every individual on the map at the precise spot where he happens to be at the moment of observation even if scale would permit a documentation of that kind.

To the explorer and visitor in their turn it seemed more economical and simpler to talk to people and record their stories than to try actually to observe their activities. The latter approach was left to the social anthropologists to develop. And so when geographers finally found that some degree of quantification was needed also for the human element we had to remain satisfied with aggregate data handed over from census takers. We were forced to accept groupings predetermined by others and very loosely related to geographical concepts. The dot map of course developed as a proxy closer to the geographers needs but since it is entirely static and does not render subgroups it remains of limited use.

Therefore, up until now first hand geographic observations of individuals and groups in action in their environment are rare. With geographic I mean observations giving the successive coordinates as time goes (cf. Chapman 1970). To this comes that what actually has been recorded and analysed starting with the individual is mostly confined to the conventional scale interval commented upon earlier: migrational moves, travel to work, shopping and recreation, farming operations. As soon as people disappear behind their doors they somehow cease to be living entities and turn at best into abstract densities per unit area.

Recently, however, some students have ventured to break through the sacred lower scale limitation and started to work seriously with the individual human being. But as is frequently the case when a reaction finally comes, it tends to jump very far in the new direction. Geographers have joined psychologists in efforts to understand certain aspects of the human mind. The endeavour to map man's perception of space around him is perfectly in order but it still leaves a large white spot to be explored: how precisely do human beings on the organismic level and viewed without the bounds of the conventional scale limitations organize their interaction and non-interaction with objects in the environment including fellow-men?

I believe it is now time to accept fully the implication of the change in outlook which perceptional geographers have brought with them. There are no longer reasonable grounds for avoiding to consider in full all actions of the individual human being. "The most important discoveries", said Lamarck, "of the laws, methods and progress of Nature have nearly always sprung from the smallest objects which she contains." For human geography it seems very reasonable to consider man to be such a central elementary particle. There is also an urgent practical consideration to take into account. For the moment no effort seems to be more important than to find ways to a just distribution of benefits and sacrifices between all human individuals. If so we should pay attention to individual behaviour as an unbroken sequence of actions irrespective of the character of the surrounding in which they take place. Scale limits are really irrelevant. Where man appears, in his house or on the Moon, something of interest to human geography happens. Generalizations must be based on concepts other than the grid size of the spatial matrix used for recording. It is also irrelevant whether the ordinary map can be used as a means of notation or not.

Two other matters are highly relevant, however. The first is the necessity to explore the connexions between the large-scale expressions of human action and what is taking place in the micro-spaces where the actors actually handle their tools and materials and cooperate face-to-face. The second is the importance of observing that the evaluation of urban structures, locations, modes of transportation, technologies and social institutions at the final end are made by people not as statistical populations but as individuals or groups in immediate interaction. Individual feelings and opinions contain the seeds of further real-world changes in the aggregate. The give and take of costs and benefits in a society are so strongly associated with minute locational acts that it is necessary for spatial analysts to apply the best magnifying glass in order to detect them. We should not, therefore, recoil at the study of individual cases. Life histories seen as sequences of actions developing in space (be it over day and week, season and year or over the total life-span) can offer a foundation on which to construct aggregative models of how individuals form bundles with one another and – in a figurative sense – with tools, buildings and pieces of land. A populated region, whatever its size, should be seen as a developing network of connected bundles, also extending out over the boundaries. On the aggregate level it must be our aim to detect and understand systems of bundles, and on the micro level we should try to find out how individuals move in and out of bundles because they want to or, perhaps more important, because they have to. The steering between bundles of various composition defines the biography of every single individual. The performance of a region on the other hand – be it a nation, a city, a neighbourhood, a work-place or a dwelling – does perhaps come out most clearly when we finally understand how it works as a gate-keeping mechanism, grouping individuals into life-curricula with different shares of wanted and unwanted events distributed in various time-patterns.

Outlines of a new frame

The suggestions I have to put forward spring from the above observations and value judgements. It would have been preferable to proceed from this point by means of demonstration in the context of actual research. Since space limitations do not permit this here, the reader is asked to accept a more speculative discussion.

Let us still take as the first fundamental assumption that the geographer sees his task as viewing the world in geometrical terms of some sort (cf. King 1969). Those who prefer multivariate statistical analysis or the application of systems concepts also begin with spatially-organized data and evaluate their findings in terms of spatial form, location and distribution. Even a purely verbal discourse, like David Harvey's (1971) analysis of the redistribution of real income in an urban system, revolves around such central concepts as accessibility and proximity, both geometrical in nature.

How do ideas with an added component of time considerations fit in with a geometrical outlook? I have pointed out that we already view the world in geometrical terms. I believe that we now must go much further along the same line and try to develop a set of concepts which makes it feasible to incorporate time with space into one unified geometrical space-time picture with full continuity in the time-direction. This means that form and process would not seem to be so essentially different as they seem today. Process takes shape as four-dimensional form. The reason for trying to move in the suggested direction is not only the hope of improved description. To man time and space are not only dimensions for viewing and analyzing the location of events, they are also in a very real sense *resources* – often scarce. This makes the space-time outlook fundamental. It is in this context that the 'packing problem' reaches its full weight and it is here we also see the importance of not leaving out any portion of man's time and surrounding space, however minute. In real life one always has to keep both dimensions in mind, for to a certain extent they are interchangeable. Furthermore, in human affairs time often makes itself felt as a more demanding dimension than do spatial constraints.

To demand a unified space-time recording of events is to ask for a great deal. The mapping tools we have today do not assist us sufficiently in this. I believe that, to produce the picture, we must – as already indicated a couple of times – develop a new kind of descriptive notation, one which is more advanced than the structural formulae of the chemist and probably closer akin to the score of the composer. The purpose must be to 'freeze' events into graphical patterns, or more likely into chains of symbols, thus converting them into a convenient form which can be viewed from various angles and which does not elude the observer while he is analyzing the message. More penetrating analysis comes after these patterns have been established, and here mathematics and statistics are surely ready to help.

Let us now for the sake of argument assume that we already possess a suitable notation. Then – being human geographers – we first of all use it for 'mapping' over a chosen part of space and a chosen period of time the complete set of unbroken space-time *paths* produced by all individuals making up the population contained in the block or passing through it. In addition we try to depict the space-time traces of all elements of known human importance in the environment, such as other organisms, tools, shelters, material and signals. Thus we have before us a sort of landscape like the bubble-chamber of the physicist.

When trying to disentangle the chaos of paths and traces inside the space-time block which the notation presents us with, it is useful to remember that human beings are always seeking to reach goals, some immediate and some more distant, some of an individualistic and some of a collective nature. This observation suggests that it must be possible to group events into coherent clusters, each cluster representing the steps necessary for movement towards each goal. Empirically – that is among past events – this classification may be difficult to carry out because a large number of goals are never attained. When looking ahead in time the task of identifying clusters is probably much easier because then we are dealing not with actual events but only with plans which always are neat and well-ordered. Let us now call the total cluster of activities, individuals and items which must participate in the reaching of some defined goal a *'project'*. The term should remind us that a project has a latent design and_ that it is moved ahead with a certain amount of 'force'. It is a project to build a New Town, to produce a ship or to write a dissertation. It is also a project to prepare and eat a meal and to clean up afterwards. Projects, conceived in this broad way, clearly form nested hierarchies. It would be difficult at this stage to define the lower and upper limits of such hierarchies. On the whole the general question of taxonomy must until further be left as an open matter.

When specifying a project one first of all has to list in time-order and quantity what goes into it in terms of *coupling with individuals and items, use of unorganized material (say air and water), and consumption of space (meaning room) and time (meaning duration)*. Place (meaning geographic location) need not appear in the specification – this is very important to note – and time appears mainly as order and duration but not, at least not always, as date, hour and second.

The specification in advance of a project, big or small, represents the largest amount of freedom an individual or a group can ever get. It only requires a knowledge of the technologies by which desired results are put together. But then, when the stage of realization arrives, real inputs have to be mobilized and many things can go wrong.

Each project would seem to try, between its beginning and end, to accommodate its parts, be they tangible or not, in the surrounding maze of free paths and open space-times left over by other projects or gained through competition with them. The important feature of a project would be sequential order; sometimes manifesting itself in terms of very fixed intervals, sometimes flexible enough for survival despite time-lags. Order in time is first of all sequence, not necessarily periodicity. *The inter-locking of projects of different life-span, up and down the hierarchy and between hierarchies, is the central problem for analysis.* This is rather analogous to the patterning of time-sharing in a computer. For the human geographer it would seem natural to choose the total available time for individuals, making up the human population, as the standard skeleton around which to build everything else.

First when we try to view the developing space-time net-work as a whole we explicitly see how much all human projects are under the strong control of a number of severe limitations. Human beings are indivisible (they can only be in one place at a time), they have restricted spatial mobility (every movement takes time from alternative actions) and they have limited life-time. If this is true for individuals it is also true for whole populations. But the same is true also for a host of other items. In fact shelters and tools have much of the same characteristics and one may even say that they form populations with specific birth- and death-configurations and age-structures. The indivisibility and mobility conditions define the possible and non possible combination of simultaneous bundles, and thus also of combinations of parts of projects. In addition, each bounded parcel of space-time can provide room only for a limited subset of individuals and items. Because of these limitations the evolving structure of nested projects can be viewed as a never ending packing process or – since human beings try to apply some control – rather budgeting process.

The world-picture which I am trying to sketch contains a further fundamental concept to which it is rather hard to give a graphic name. Tentatively we might call it 'budget-space', indicating that it has to do with allocating different parts of a limited whole to specific purposes. Now for a specified project to become realized in a real-world situation, both when it is controlled by conscious human action and when it is moved forward by inner biological forces, it must be able to fit into 'cells' in the budget-space within reach. What is within reach depends first of all on how much time for movement the project can accommodate without being disrupted. In small-scale human terms an open cell could be time free for some activity; in building terms it could be, for example, an empty room. When an activity has started or a room has become occupied these cells become closed for some duration of time. Other projects looking for the same locations in space and time have to await their turn or go elsewhere.

This process of fitting projects into budget-space is an intriguing kind of applied topology, perhaps like laying a jig-saw puzzle with rubber-pieces. There must always be some conditions of internal connectivity within the project which cannot be violated if it is to survive distortion. Weaker components of a project (weaker in a relative sense) have to search surrounding space-time for cells which are empty or at least have a still weaker content which will admit entrance to the cell. Stronger components (again in a relative sense) may be able to force themselves into comfortable cells, but when so doing they frequently start spread-effects affecting other projects, likewise seeking to survive and to develop their interests somewhere in total budget-space. Distance between components is important in this context, as already indicated, because the necessary movements take time. If too much time is consumed by movement it may not be possible to maintain the order within the project necessary for its survival. The order-breaking effects of movement because of time-consumption is perhaps a more important constraint than the constraint of movement *per se*. The advantage of the car, to take a practical example from everyday life, may first of all lie in its order-preserving capacity for the projects in which the driver is involved, compared with the difficulties he would meet if he were constrained by time-tables. It is well known that lowering fares has a very small influence in attracting car-drivers to use public transport. Telecommunication has in the same way contributed to order-preserving in information-handling projects. Still, because such projects to a large extent seem to be built up as mixtures of easily-transmitted information and more complicated information must be carried personally, a project as a whole may remain sensitive to ordinary costs of transportation. The question of how, in the long run, telecommunications will affect the location of information-dependent activities cannot easily be estimated. The critical factor is how the various parts of an information project hang together and what kind of cells in budget-space it needs to occupy for its successful completion. It is clear from the above cases that when competing for budget-space projects are susceptible to many sorts of accidents. They become held up by each other, get crippled, die out completely before a full program is finished, or have to see the programs redefined. It is also clear that if we can arrange a description in the terms which are suggested above then we will have a frame for safe deductive reasoning. We are dealing with the partition of a whole into pieces in the way one does in accounting and we can thus see how arrangements in one corner necessarily affect the situation in another.

An important consequence of looking at things in this way is that one accepts spatial location to be strongly determined by the sequence of events in time. It is a frequent experience in practical planning that a real choice between alternative locations is a rare situation. This is so, not because alternatives in geographical space are not available, but because deadlines in time and links with other projects do not admit a choice. The answer to this has been to try to plan ahead longer and longer and, recently, to back up these long-term plans with broader studies of the future. In my view the planning of spatial location will not work well until we have a better understanding of how projects as wholes must be accommodated in a space-time budget-space. Nor can we understand how various arrangements of a political, technical or locational nature affect individuals in a population unless we can estimate how their private budget-spaces become widened or circumscribed by the proposed actions. The out-come of competition between projects determine how sequences of events fall out over individuals in the population concerned.

The maps we use today bring out spatial (usually only two-dimensional) cross-sections through projects which are in various phases of their progress. How far momentarily-observed relations, measured in terms of distances, configurations and densities can reveal the counterpoint structure over time must remain a matter for investigation. That some of them emerge is clear from Curry's discussions of ergodic processes (1967) and of entropy relations in trading systems (1971). In a way it is an ironic circumstance that most other quantitative techniques so far applied in human geography seem to be best fitted to deal with the real-world situation of an old-fashioned, stable rural environment where friction of distance is immensely high and the projects related to human action are on the whole strongly repetitive and restricted to compact space-time "bubbles" which are elongated in time but very narrow in space. Central-place theory belongs fundamentally to this world and so does my own conceptualization of diffusion. In the urban-industrial world the competitive accommodation of projects is much more complex and above all much more spread out over terrestrial space. But still these projects must be well ordered in time, since order in time is the fundamental condition for survival of a project. This is an observation we ought to keep in mind when planning future investigations in human geography.

To accommodate a project in budget-space involves much more than simply controlling the relative space-time locations of components. The cells in budget-space have not only co-ordinates in space and time but there are also various entrance conditions attached to them. This political aspect will be taken up later. They have in addition variable internal characteristics such as spatial volume, duration, and arrangements for the accommodation of lower-order projects. People as resources have different professional competence, plants different nourishment values and so on. When a family is looking for a home, a firm for office premises and a general for a battle-field, the internal attributes of possible structures are of crucial importance. I believe that on many occasions considerations of relative location play a part only within broad tolerance limits, compared to the decisive effects generated by internal attributes of possible locations. It seems that most human decision-makers have such a vague idea of the space around them that the outside world must impinge on them personally before they take the situation seriously. Only then do they start to look for solutions, but when testing what is acceptable or not they usually make decisions on the basis of immediate impressions rather than trying to search a wider territory. Let us note the possibility that the internal attributes of buildings can explain location of people and activities in urban zones equally well as relative distances between interacting units. This is not a novel observation, of course, but it is perhaps underestimated.

It is thus as an aggregated result of the choice of cells in budget-space that unlike things come to be mixed in space and enter into proximity relations without otherwise interacting with each other, in any direct, functional sense. Out of the mix come external effects, good or bad depending on one's viewpoint.

Given this general framework it should be feasible to identify classes of situations which stand out as critical in the process of continuous rearrangements. Let me point out some important mechanisms. The first, which has been empirically investigated to some extent, is responsible for starting chains of moves and could be labelled "outgrowth". By this I imply that some cell in budget-space has accumulated projects or parts of projects up to a point beyond its holding capacity. In many instances a certain amount of elasticity would prevent an explosion from taking place, but sooner or later a colonization of surrounding budget-space is unavoidable. Out-migration from the family of grown-up children (Hägerstrand 1969) represents a model case. Common also is the 'planting out' of subsidiaries or annexes by growing firms and institutions as an intermediate stage before complete relocation (Améen and Erlandsson 1969 and 1971). The out-growth process is interesting from the point of view of location theory, for the surrounding territory is seldom an empty and open budget-space. Other projects are already there. Thus the set of available cells is somehow limited in number. Add to this the fact that opportunities may change quite quickly over time because of competition, which means that what is open for choice is very much dependent on the point in time. Workable opportunities are really only those which are known to the decision-maker, which have acceptable internal characteristics and which permit entrance at the right moment. Therefore, the internal circumstances within cells between which interchange may take place, as well as suitable co-incidence in time, become very decisive factors for the outcome in space. It is not surprising that spatial patterns on maps so often appear to be random (Curry 1964). Nevertheless it is altogether possible that projects going on in the spatial mix have quite deterministic configurations when analyzed in terms of the available budget-space. However, the empirical problems related to the actual investigation of these matters are of course extremely difficult to solve.

Conditions for entrance were mentioned already as a constraint.

Here again, we meet a phenomenon of fundamental importance in the budgeting process. Budget-space, organized by man, is covered by domains in which various forms of conducts can be prescribed by someone. The boundaries are in certain cases legally protected to guarantee their stability. In other cases the conditions for entrance and conduct are culturally determined. Formal and informal domains with variable life-time form together an extremely complicated network of super- and subordinated entities, designed to protect sections of space-time against invasion in order to screen off on-going projects from unwanted influences, and to give reserve budget-space for the accommodation of planned or otherwise expected future events.

Power and space-time is thus an immense area for research in human geography which for some reason has hardly been touched. Perhaps one of the causes for this neglect has again been our traditional tendency to concentrate on the visible elements of the landscape. Most political boundaries are not directly visible. Such power over space and time which can be transferred between individuals and coalitions by trading on a land market, is to some extent acknowledged or at least taken into account by tacit assumptions, but very little work has been done on domain-systems at the scale between the plot and the nation-state. At the other side of the limit almost nothing is known about how corresponding informal structures function in the micro-spaces of homes, workplaces, playgrounds and streets.

The whole complicated question of sectoral (vertical) versus areal (horizontal) planning and control is involved in this. The same pattern of conflicts seem to emerge at all scales from, for example, the clinic of a hospital (in which administrators and doctors have different views) to the national territory (in which central government departments and local governments have different views). I have been rather deeply involved in the redivision of my country into new local government areas and I know from personal experience that the most unexpected complications emerge here of far-reaching consequence for location of industry, public investment and undertakings concerned with environmental control. Few studies exist of how high-order decisions become modulated when they are filtered down or how influences work the other way round.

What is available indicates a fertile and important field of research (Anderson 1970). Some human geographers have believed that political scientists have something to say in this regard but this does not seem to be the case. Their geometrical assumptions are far too weak to be of use in the kind of theoretical structure which I advocate here. The same can be said about the behavioural scientists in general. They mostly deal with direct power relations between people (Kelvin 1969) but that is knowledge about the actors, not about the conditions for action imposed by arrangements on the stage. What is in fact called for is a new type of political geography, dealing with power in space-time terms of considerable conceptual precision. Power relations are of such immense importance for the understanding of how projects compete in available budget-space that a well-conceived political geography could well develop into the core of human geography.

It in interesting to note that it is just by way of political geography of the kind which I have in mind that we come into closest contact with physical geography. Chorley and Kennedy (1971, 298) have recently pointed out that we are entering a period when man, "organized in ever more effective decision-making groups, is increasingly able (at least in theory) to exploit his growing knowledge of the nature and operation of natural process-response systems so as to be able to intervene in them to exert an influence which will modify their operation in a planned and predictible manner, beneficial to man in the widest sense ..." The critical concept here from the point of view of human geography is the concept of 'decision-making groups'. Such groups do not exist in a vacuum. They have, and will probably always have, some sort of given or achieved territorial competence (which does not necessarily mean that they control everything inside their territories) with subordinate domains inside and competing domains of the same rank outside. The physical and/or ecological situation may well define a 'rational' system of domains, adjusted to the operation of the natural processes to be controlled. But man also has other projects going on at the same time and in the same space, including the private projects of the population, and it is not likely that the space-time division of power will turn out to be an easy business to settle in order to fit only some selected natural process. Conflicts are unavoidable.

Budget-space of mankind is a battleground between projects. The scene, taken in the aggregate, is dominated by a Darwinian kind of trial and error process. Mankind is now beginning, at least in the opinion of many, to be threatened by the sheer size and complexity of its own projects. It is certainly time to sit down and think quietly about what human geographers can contribute to the understanding of the situation. The competence we have to invest lies in our experience in interpreting the world in geometrical terms. If we try to build out this experience towards a space-time geometry with an emphasis laid upon the fitting of projects into limited budget spaces then we may arrive at a conceptual system of greater strength. The combination of geometry and accounting procedures may open the way for future-oriented deductive reasoning of a kind which has not been available before.

References

ABLER, R., ADAMS, J.S. and GOULD, P. (1971) *Spatial Organization: The Geographer's View of the World*; (Prentice-Hall, Englewood Cliffs).

AMEEN, L. and ERLANDSSON, U. (1971) *Institutionella tillväxtförlopp i det urbana rummet*; Urbaniseringsprocessen 33, (1969), 43, (Lund). (mimeographed research reports).

ANDERSSON, L. (1971) Decision-making and spatial changes; *Urban and Regional Planning, London Papers in Regional Science 2*. (Ed. A.G. Wilson), (London).

BUNGE, W. (1969) Simplicity; *Geographical Analysis*, 1, 388-91.

CHAPMAN, M.T. (1970) *Population movement in tribal society: the case of DuiDui and Pichahila, British Solomon Islands*; (University of Washington, Seattle).

CHORLEY, R.J. and KENNEDY, B.A. (1971) *Physical Geography. A Systems Approach*; (Prentice-Hall, London).

CURRY, L. (1964) The random spatial economy: An exploration in settlement theory; *Annals of the Association of American Geographers* 54, 138-46.

CURRY, L. (1967) Central places in the random spatial economy; *Journal of Regional Science* 7, 217-38.

CURRY, L. (1971) Geographical specialisation and trade; *Urban and Regional Planning, London Papers in Regional Science 2* (Ed. A.G. Wilson), (London).

HARVEY, D. (1971) Social processes, spatial form and the redistribution of real income in an urban system, Ch. 13 in

Chisholm, M., Frey, A.E. and Haggett, P. (Eds.) *Regional Forecasting*, (Butterworth, London).

HUDSON, F.C. (1969) Location theory for rural settlement; *Annals of the Association of American Geographers* 59, 365-81.

HÄGERSTRAND, T. (1969) On the definition of migration; *Scandinavian Population Studies* I, (Helsinki).

KELVIN, P. (1969) *The Bases of Social Behaviour: An Approach in Terms of order and values*; (London).

KING, L. (1969) The analysis of spatial form and its relation to geographic theory; *Annals of the Association of American Geographers* 59, 572-95.

5 · New geography as general spatial systems theory—old social physics writ large?

WILLIAM WARNTZ

Introduction

In 1958 John Q. Stewart and this author (1958a) identified a "macro-geography" of human phenomena including the recognition of number of people, distance and time as basic categories or dimensions. It was argued that only from the "thorough appreciation of these brute physical factors can grow a correct and fertile treatment of the loftier human characteristics, which, too, must be included in their turn." It was added that "perhaps much of the lag in the development of social science stems from the impatient attempts to understand the latter more difficult problems before time and space regularities are understood – and to attempt to analyze 'social change' before understanding social equilibria." In support of these ideas wide varieties of empirical evidence relating to social and economic phenomena were examined and demonstrated to be highly correlated in their spatial patterns in the United States and elsewhere to measures of derived quantities (relating numbers of people, incomes, distances and time) and isomorphic in their mathematical formulations to those in "field quantity theory" in the physical sciences generally (Stewart and Warntz 1958b).

Care was taken to point out that the size of the area as portrayed on a map does not indicate whether the approach be microscopic or macroscopic. The mere assembling of more and more area and increase in detail in itself involves no shift from micro to macroscopic. A heightening of the level of abstraction is the significant thing, and insistence on the functional consistency and "organic unity" of the whole, a recognition that no part of a *true system* can be understood thoroughly without reference to the whole. And, in support of the general idea of (spatial) "potentials of population" (and related concepts), the idea was advanced that it was a sufficiently abstract and subtle measure of position (such a measure having long eluded geographers) to help make possible the development of a macro-geography capable of producing generalizations about *space-occupying* systems.

The authors were fully prepared to receive the criticisms that followed this excursion into "social physics". Not unexpectedly, the comments ranged from purely emotional outcries against the "inappropriateness" of the measures from a so-called humanist point of view to valid scientific questions and comments about the nature of the constants involved in the published equations, the interpretations of the statistical measures, and the "meanings" of the results. In addition many recommendations about additional research topics were received.

From the above studies, as well as others, a nexus of patterns of relationships emerged, mathematically similar to primitive ones already well recognized and established in the physical sciences. From this Warntz (1965) went on to study spatial and temporal patterns of income distribution in the United States from colonial times to 1959 and to make explicit the many ties with conceptual structures, for example in thermodynamics.

Apparently at that early date it could not be stressed too often that the similarity in structure of patterns derived from no *a priori* determination. Rather, this result was obtained because both kinds of abstractions are logically related to each other. No equations were lifted bodily from one discipline to be used as a rigid mold into which data from another domain were to be crammed. Rather, if anything was transported into the discipline of geography it was a mental set and the attitude that one should be prepared to accept what the data reveal when subjected to testing — a willingness also to understand, however, the results may never be taken apart from the design of experiment. Similarities in structures, particularly in spatial configurations, revealed themselves among data often taken to be dissimilar when classified by their non-spatial properties.

Elihu Fein (1970) has noted this similarity of structures and has extended and enlarged upon the work of Stewart and Warntz and especially Warntz's concept of Income Fronts. His contribution is to be reviewed later in this essay. It suffices now to note that Fein insists that "the conclusion is not that people act like molecules. This simplistic interpretation misses the essence of the argument ..... Rather, it may be asserted that a similar organization of knowledge can apply to two different classes of aggregate behaviour."

As we perceive molecules and people in certain well-defined circumstances to behave as aggregates, we will be satisfied to use common functional cause-effect relationships. That this structure — these relationships — is the set of abstractions by which we originally understood the physical behaviour of a gas does not limit it phenomenologically. On the contrary, the fundamental postulate here is that structures, saying whatever they do about our perceptions, may be quite general.

This generality is, of course, well attested, too, in the recent and ongoing development of "General Systems Theory" led, of course, by Ludwig von Bertalanffy (1968). He emphasizes that a general systems theory is not an investigation of hazy and superficial analogies having little value because differences as well as similarities can always be found among phenomena. (Particularly is this true at classificatory levels.) Rather, Bertalanffy stresses, too, that isomorphism which is a consequence of the fact that "in some respects corresponding abstractions and conceptual models can be applied to different phenomena."

Indeed, the early macrogeography of social and economic phenomena stemming from social physics has found its intellectual base and the means of its extension through General Systems Theory. But, so, too, have other branches of geography. Chorley (1962) has, for example, given expression to this for geomorphology. Berry (1964) has written of "Cities as Systems within Systems of Cities" and Woldenberg and Berry (1967) have considered rivers and cities. Other examples abound. A general statement and a useful summary is to be found in Ackerman's (1965) statement for the National Academy of Science — National Research Council of the United States concerning the science of geography.

In addition, attention might be called to the series of papers recently issued (1966-1971) from the then existing small group of research-oriented theoretical geographers. In assaying the recent literature in geography relevant to General Systems Theory they became impressed with the need to make explicit the significant geometrical and topological properties of surfaces taken generally if spatial structure and spatial process were to be understood as general phenomena in neutral terms apart from either the so called physical or socio-economic non spatial properties, for example, attributed to the phenomena. A first paper on the general properties of surfaces (Warntz and Woldenberg 1967) established the concepts from which a long series of other papers derived. Increasingly it was recognized that the distinctions among the various systematic branches of spatial study diminish at the theoretical level as common spatial properties are recognized and as spatial solutions increasingly cut across the traditional subject matter of geography. During the relatively short period this group was allowed to work unhampered by the rapidly mounting chaos surrounding it, it came to see clearly the vision of Geography as General Spatial Systems Theory and to appreciate the significance of relating the level of abstraction required to the analysis to be undertaken. The analysis of geographical patterns regarding the significance of population distributions (for example – to relate to what follows in the rest of this essay) requires various levels of generalization and abstraction as reformulation of hypotheses suggest it. By increasing the level of abstraction is meant not introducing additional vague, nebulous, or confusing ideas, but rather, increasing the information content of measures and symbols. For example, in the particular terminology of the kind of work to be described below, it can be demonstrated that the dot, density, potential, energy, and action levels for consideration of populations follow each other in a normal, natural, sequence, each in an ascending order of abstraction and information generalization in an appropriate gamut ranging from microscopic through macroscopic consideration of population systems.

To exhibit this and to call attention to further developments in the earlier work in the macrogeography of population and income distribution as this is related to emerging concepts in Geography as General Spatial Systems Theory, the following is offered. The format follows that of the earlier work, but the data and their analysis, are new.

The geographical distribution of income in the United States (1967-68)

In the conterminous United States, as of January 1, 1968, an estimated 198,198,500 persons were distributed over the 2,968,747 square miles of area and shared approximately 534,336,000,000 dollars per annum as their disposable personal income as measured for the preceding year.

Complete knowledge of the microgeographic distribution of this population and its income would include information on the precise geographical location assignable to each individual and on his specific income. Here the emphasis on the unique and particular, with the dot map as the most suitable means of presentation, may be useful for many purposes. But, at this level, analysis is severely limited. Hence, increased macroscopy and a higher level of abstraction is required.

From the values of these primitives of analysis or dimensions, i.e., population and area (or more properly distance, since area is defined by distance squared), together with income, a number of derived quantities can be computed which facilitate the investigation of spatial patterns of population and income distribution in the United States. Time is another such dimension or primitive of analysis but will not, in the first instance, be considered here.

Let these basic dimensions be identified by the following symbols:

P = Population;
I = Income;
r = Distance.

Then A = Area (in units of r squared).

With number of people, dollars, and miles as convenient units of measurement, the following derived quantities are defined and used subsequently:

$$T = \text{Per Capita Income} = \frac{I}{P};$$

$$G = \text{Population Density} = \frac{P}{A};$$

$$D = \text{Income Density} = \frac{I}{A} \text{ and alternatively } \frac{PT}{A} \text{ or GT};$$

$$U = \text{Income Potential at a point} = \int \frac{1}{r} D dA \text{; and}$$

$$V = \text{Population Potential at a point} = \int \frac{1}{r} G dA$$

Where the distance is taken from each infinitesimal element of area to the given point.

The values $\int(1/r)DdA$ and $\int(1/r)GdA$ cannot be computed directly, but a mechanical integration based on summations can yield approximations of required values that will be as nearly accurate as desired. Whether computations are performed longhand or by high speed electronic digital computers, the arithmetic procedure remains essentially the same. For income potential, for example,

$$U_i = \frac{I_1}{r_{i1}} + \frac{I_2}{r_{i2}} + \ldots + \frac{I_n}{r_{in}} + \sum_{j=1}^{n} \frac{I_j}{r_{ij}}$$

with an appropriate step taken for $i = j$

An obvious but important definitional truism is:

DA = PT

Total income is represented by both sides of this identity. The left hand side considers the areal distribution while the right hand side concerns the distribution of the income among the population.

Now if the population of the conterminous United States had been evenly distributed over its area, each spare mile would have contained nearly 67 persons. An even distribution of the income among the population would have resulted in 2696 dollars per person. And, given these two conditions, the income density everywhere would have been approximately 180,900 dollars per square mile.

These conditions manifestly did not exist. Table 5.1 shows that if the above values of G, T and D are computed as averages on a *state* basis, population densities range from the 3.6 per square mile in Wyoming to 938.9 in New Jersey, per capita incomes varied from 1739 dollars per person in Mississippi to 3292 in Connecticut, and income densities exhibit a low of 8370 dollars per square mile in Wyoming and a high of 2,908,400 in New Jersey. The ranges will be seen to be even greater when county values are considered subsequently.

Table 5.1 displays the gamut of state values for each of these derived quantities. Although state areas and populations vary greatly in size, and boundaries are admittedly arbitrary for many purposes of analysis (but not all), the incontestable and patently obvious fact exists that considerable geographical variation in these phenomena occurs. Certainly no one would dispute this. But, what more can be learned about these variations and what relationships discerned?

(a) Income potential and income density

In fig 5.1 is shown, as of January 1, 1968, the Income Potential Map for the United States. Maps of this sort have manifold applications and their properties are becoming well known. Here it will be stressed that from the microgeographic discrete distribution of population and income, a macrogeographic continuous distribution has been created. Income potential in dollars per mile is a spatially continuous macrogeographic variable and the value of the intensity of potential at any point is a scalar in an areal continuum of such quantities. Thus income potential is a "field quantity", analogous to gravitational potential, and has similar characteristics such as mathematically derivable intensities, gradients, forces, and energies, all of which have been used in other analyses.

The scatter diagram of fig. 5.2 shows the high degree of correlation between income density and income potential by states in 1967. Both scales are logarithmic and the relationship is clearly linear in the logs. The simple formula describing the relationship is of the following form:

$$D = kU^w$$

where D and U are defined as above with k as the factor of proportionality and w as the exponent of the income potential.

When D for a state is taken in dollars per square mile, and the representative U for a state is taken in dollars per mile, a least-squares solution yields the following:

$$\log D = 21.966 + 3.02257 \log U$$

Table 5.1 Areas, populations, incomes, and derived quantities – by states in conterminous United States, 1967

No. State	I	II	III	IV	V	VI	VII
	A	P	I	G	D	T	U
1. Washington	66.8	3128	9189	46.83	137.56	2938	494
2. Oregon	96.3	1994	5225	12.44	54.26	2620	484
3. California	156.7	19468	60022	124.24	383.04	3083	825
4. Nevada	109.8	478	1393	4.35	12.69	2917	557
5. Idaho	82.8	705	1585	8.51	19.14	2249	488
6. Utah	82.3	1037	2362	12.60	28.70	2278	543
7. Arizona	113.6	1676	3822	14.75	33.64	2280	517
8. New Mexico	121.5	1060	2209	8.72	18.18	2084	541
9. Colorado	103.9	2038	5286	19.62	50.87	2593	603
10. Wyoming	97.5	361	816	3.60	8.37	2325	534
11. Montana	145.9	725	1728	4.97	11.84	2385	473
12. North Dakota	70.1	647	1474	9.23	21.03	2279	584
13. South Dakota	76.5	709	1577	9.27	20.61	2225	657
14. Nebraska	76.7	1489	3946	19.41	51.45	2651	719
15. Kansas	82.1	2293	6084	27.93	74.10	2653	777
16. Oklahoma	69.0	2494	5758	36.14	83.45	2308	786
17. Texas	263.5	10985	25744	41.69	97.70	2344	679
18. Louisiana	45.2	3671	7961	81.22	176.13	2169	792
19. Arkansas	52.7	1958	3738	37.15	70.93	1909	873
20. Missouri	69.2	4368	12041	66.01	174.00	2535	1043
21. Iowa	56.0	2808	7805	50.14	139.38	2779	917
22. Minnesota	80.0	3638	9659	45.48	120.74	2544	792
23. Wisconsin	54.7	4272	11316	78.10	206.87	2549	1055
24. Illinois	55.9	10897	34955	194.94	625.31	3208	1443
25. Mississippi	47.2	2333	4058	49.43	85.97	1739	871
26. Alabama	51.1	3538	6835	69.24	133.76	1937	944
27. Tennessee	41.8	3885	8122	92.94	194.31	2091	1110
28. Kentucky	39.9	3172	6789	79.50	170.15	2141	1236
29. Indiana	36.2	5019	14210	138.70	392.54	2831	1446
30. Michigan	57.0	8511	25375	149.32	445.18	2932	1294
31. Ohio	41.0	10662	29166	260.00	711.37	2736	1598
32. West Virginia	24.1	1766	3725	73.28	155.21	2109	1360
33. Georgia	58.5	4465	10004	76.32	171.01	2241	941
34. Florida	54.3	6166	14662	113.55	270.02	2378	819
35. South Carolina	30.3	2630	5066	86.79	167.19	1925	995
36. North Carolina	49.1	4994	10668	101.71	217.27	2136	1124
37. Virginia	39.9	4577	10914	114.71	273.53	2384	1344
38. Maryland & D.C.	9.9	4538	13727	457.40	1386.57	3031	1858

Table 5.1. (Cont.)

No. State	I	II	III	IV	V	VI	VII
	A	P	I	G	D	T	Ụ
39. Delaware	2.0	521	1583	260.50	791.50	3038	1737
40. New Jersey	7.5	7042	21813	938.90	2908.40	3098	2346
41. Pennsylvania	45.0	11717	31539	260.40	700.87	2692	1807
42. New York	47.3	18223	56987	385.24	1132.98	3127	1619
43. Connecticut	4.9	2943	9687	600.61	1964.94	3292	1940
44. Rhode Island	1.1	903	2516	820.91	2287.27	2786	1711
45. Massachusetts	7.9	5458	16127	690.90	2041.39	2955	1700
46. Vermont	9.3	404	1013	43.44	108.92	2507	1154
47. New Hampshire	9.0	671	1781	74.56	197.89	2655	1145
48. Maine	31.0	985	2269	31.77	73.19	2303	734

I Area (in miles^2) $\times 10^{-3}$
II Population (in persons) $\times 10^{-3}$
III Income (in dollars) $\times 10^{-6}$
IV Population density (in persons/mi.^{2})
V Income density (in \$/mi.^{2}) $\times 10^{-3}$
VI Income per capita (in \$/person)
VII Income potential in (\$/mi.) $\times 10^{-6}$ (Contribution to self using uniform circular assumption)

or when stated in the power form with the same units, this can be approximated by:

$$D = 01.08 \times 10^{-22} U^{3.02}$$

The coefficient of correlation is 0.954 and the association may be deemed highly significant. (For the number of statistical degrees of freedom here obtaining, the coefficient of correlation for the 1% level of significance is 0.37).

The above result was empirically derived from one year only. As will be shown, however, similar observations for other years reveal the same general relationships. Deviations do exist, as shown in fig. 5.2, and the deviations in each particular portion of the curve are undoubtedly of considerable interest in themselves. But, it can be demonstrated that income potential must be a *smooth* continuous surface. Therefore the high correlation between income potential and income density permits the assumption of a spatially continuous

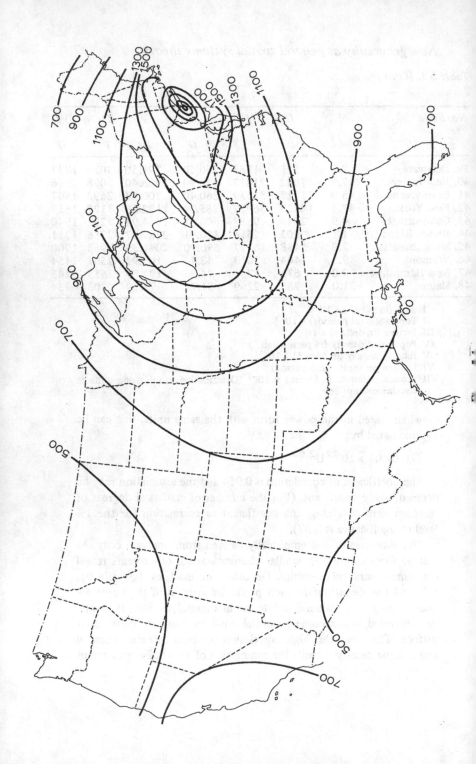

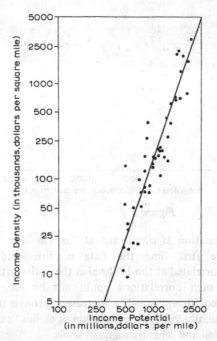

Figure 5.2

smoothed income density surface. However, this assumption is only a *first approximation* and only holds within the limits permitted by this high correlation and the probability level employed. (The usage of *smooth* and *smoothed* is deliberate and the distinction is important.)

(b) Income potential and per capita income regions

Figure 5.3 shows the first exploratory scatter diagram for the same 48 states for per capita incomes plotted against income potentials. The lack of a very high correlation is obvious. A least-squares solution to the logs of the values produces a coefficient of correlation of 0.646 and when the value for a coefficient of correlation at the 1% level of significance again is 0.37, then one cannot accept the notion of an overwhelmingly important relation between the two variables at first glance. It must be noted, however,

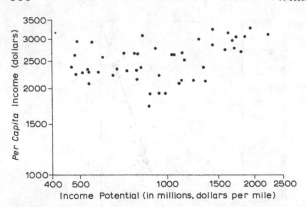

Figure 5.3

that this correlation is significant at the 1% level and that this represents the first time the data of this kind are deemed significantly correlated at the 1% level at the entire national level. At earlier dates such correlations could not be judged significant although the coefficient steadily increased through time as will be shown subsequently in the recapitulation of earlier "experiments" at the national level and separate regional levels.

Study of the deviations here, however, is extremely rewarding. The solution indicated above was based on a formula of the following form:

$$T = zU^x$$

When T, the per capita income, was taken in dollars per person and U, the income potential was again taken in dollars per mile, the following resulted:

$$\log T = 2.084 + 0.14636 \log U$$

When stated in the power form, the equation is approximately:

$$T = 121 \, U^{0.15}$$

The deviations here, unlike those found for the correlation between income density and income potential, are not only much larger in terms of logs but also quite importantly do constitute certain definite geographical patterns to be described subsequently.

Thus, they differed from those of the first correlation in two noteworthy ways.

To study the nature and importance of these deviations, a tedious and laborious, but straightforward process was begun based on the conventional theory of outliers. The greatest deviation in terms of logs was eliminated. In this case it was the low value for Mississippi (state number 25). Correlations were again done minus this elimination. Of course, the coefficient increased very slightly and the constants changed very slightly. (It is to be noted that this and all other correlations referred immediately below are based on weighted regressions with each state's representation proportional to its population). Next, South Carolina was also eliminated. Then, the high value of Washington, the low of Arkansas, and so on, were included in the cumulative elimination. At various stages in this cumulative elimination process, the least-squares solution was reworked. Of course, the effective coefficient of correlation steadily grew larger even when corrected for the losses in statistical degrees of freedom. The factor of proportionality and the exponent of potential also changed individually and slowly, sometimes increasing and sometimes decreasing. Finally, when the coefficient of correlation became comparable to that of the income density-income potential correlation, the successive cumulative eliminations resulted in 33 remaining states for which the following weighted regression equation applies.

$$\log T' = 1.523 + 0.2116 \log U.$$

The coefficient of correlation is 0.931, a figure in satisfactory agreement with the 0.954 of the overall income density and income potential relationship. The thirty-three participating T' states form a "main sequence" extending from Maine (state number 48, see Table 5.1) to Arizona (7). The states included are numbered 5-17, 20-24, 29-31, 34, and 38-48. These states are denoted by the large dots on fig. 5.4 included between lines B and C. Line A, of course, is the line of best fit for these 33 states defined by the equation next above. In the power form (and again with much of the fictitious accuracy eliminated) the equation is approximately:

$$T' = 33.34 \, U^{0.21}$$

the units of T' being as noted above, dollars per person, and U dollars per mile.

But, what of the eliminated deviating states? They group themselves into two distinct geographical regions corresponding on the one hand to the statistical class of values that were too high, and on the other to those that were too low. High values of per capita income, T'', are found in the Western part of the United States, of "The Far West" as it is called, and include states 1-4. The low values, T'' are found to constitute a compact region of states 18, 19, 25-28, 32, 33, and 35-37 in the southeastern part of the United States or, as it has been called traditionally, "The South".

The T'' or high value states all lie above line *B* in fig. 5.4 with the dotted line *E* fitted to their distribution. The low-value states all lie below line *C* on the chart with the dotted line *D* fitted to their distribution. Note again the very significant coincidence of statistical classes and geographical regions.

Heretofore line *B* and *C* have been noted only as graphical boundaries to the distribution of the 33 main sequence T' states. Now it should be added that these lines, which, it will be noted, are not quite equally spaced, are not just arbitrarily drawn. Rather, they are placed, respectively, at +1.67 and −1.67 standard errors of

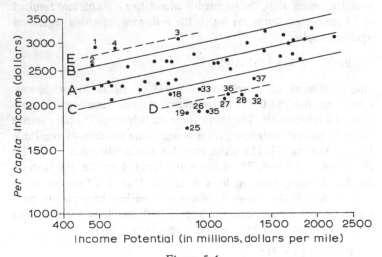

Figure 5.4

estimate about the main regression line, A. They thus give a measure of the degree of "isolation" of these subsystems or regions in the general system. The standard error of estimate in logs for the 33 state correlation is 0.0294.

When a weighted least-squares solution was obtained separately for the 11 low valued states of the T''' region the following equation resulted:

$$T''' = 26.31 \, U^{0.21}$$

For the 4 high values states of the T'' region:

$$T'' = 40.55 \, U^{0.21}$$

For each of these two regions the correlation coefficient exceeded 0.80 when both constants were determined by the data. The separately and independently determined x values were found to be 0.26 and 0.23 respectively and upon testing revealed to be not significantly different from a universe value of 0.21. On the other hand, the z values all test as significantly different from each other whether determined freely or with the "fixed" slope of 0.21. Thus, three regions can be recognized. The three values for z represent scale factors in what can be regarded as spatially defined subsystems in a general open system relating to the flow of information, ideas, capital, commodities, payments, and people. It will be very instructive to consider income and other social and economic phenomena in terms of the sequence of adjustments in regional organizations for growth and for decay and feedbacks affecting the forms of the dynamic allometric system equilibrium among number of people, distance, time, and income. To this end the income density — income potential and per capita income — income potential relationship have been examined for earlier periods and the results of these investigations will be presented subsequently.

For the time being, however, for 1967, it can be observed for the general equation

$$T = zU^x$$

the same value of x, the exponent of the income potential with which per capita income varies, can be applied to all three per capita income regions. This value is 0.21, that is, approximately the fifth root. The values of z, the factor of proportionality, constitute the significant difference among the three regions with $z' = 33.34$, $z'' = 40.55$, and $z''' = 26.31$ for the units used. Figure 5.5 shows these three per capita income regions.

The correlation between income density and income potential showed that, as a first approximation (for the levels of significance involved), no separate income density regions exist. However, the similarly significant first approximation for the per capita income and income potential correlation reveals three per capita income regions. It should be borne in mind that the ensuing analysis and subsequent suggestions are based upon these first approximations.

(c) Population "weights"

The values of z may also be regarded as indicating population "weights". If $z' = 33.34$ be taken as a basis, unity, then the average individual in the Main Sequence or T' region can be regarded as one standard person. The foregoing correlations remain unchanged. Then, the population of the states of the Far West or T'' region can be regarded as "increased" from the population of actual persons to a population of standard persons by a factor of 1.216. If per capita incomes are then computed as dollars per standard person, they all fall into general agreement with line A in fig. 5.4. This value, 1.216 is of course the ratio of Z'' (40.55) to z' (33.34).

Similarly, the ratio for the states in the South yields 0.789 as the factor to give the population in terms of equivalent standard persons. When this adjusted smaller standard population is divided into incomes, the resulting higher value of dollars per standard person is also in agreement with line A.

By the above transformation the three separate per capita income regions disappear as do the boundary discontinuities of the per capita income surface. But, regional weighting of population and discontinuous surfaces emerge as their exact counterparts. This transformation is extremely interesting because the weights thus deducted are in substantial agreement with those found empirically for general interaction of population at a distance and such special interactions as the movement of college students, flow of bank checks, telephone messages and passenger flows.

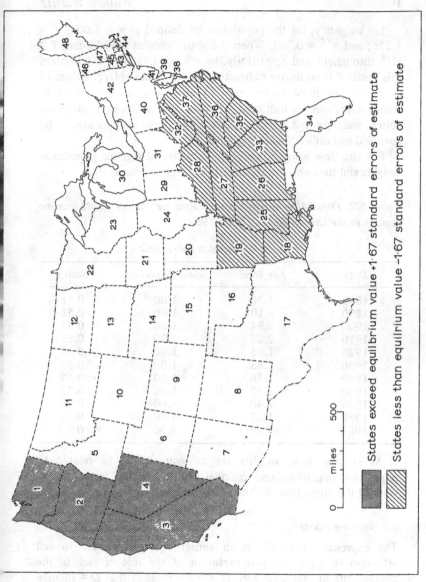

Figure 5.5

States exceed equilibrium value +1·67 standard errors of estimate

States less than equilrium value −1·67 standard errors of estimate

0   500
miles

Let weights y, for the population be defined as $y' = 1.000$; $y'' = 1.216$; and $y''' = 0.789$. When the above weights are adopted, $T = zU^X$ throughout and empirically, here $T = 33.34\ U^{0.21}$. To obtain this result T is no longer defined as I/P but rather as I/yP. Values of G can be modified accordingly to give population densities in standard persons per unit area, as yG. The magnitudes involved for actual and standard populations for the three regions are to be specified and utilized in analysis later.

For the time being however, a summary of regional population weights and their changes through times is given below.

Table 5.2. Population weights as averages for the per capita income regions in the United States.1880 - 1967

Population weights by regions

Date	Far West	Main Sequence	South
1880	3.30	1.00	0.51
1890	3.10	1.00	0.51
1900	2.94	1.00	0.52
1910	2.27	1.00	0.55
1920	1.95	1.00	0.58
1930	1.83	1.00	0.59
1940	1.78	1.00	0.59
1950	1.52	1.00	0.71
1956	1.40	1.00	0.76
1959	1.37	1.00	0.77
1967	1.22	1.00	0.79

As before, these weights are subsequently to be related to population migrations and regional capital transfers in the framework of the dissipation of "Income Fronts".

(d) "Income fronts"

The expression $DA = PT$ given earlier is repeated here to call attention to a possible interpretation of the role of each of the components in that equation. It will be recalled that D = Income Density, A = Area, P = Total Population, and T = Per Capita Income. There appears to be a similarity between the above equation and the equation of state for the perfect gas within the macroscopic gas laws of Boyle and Gay-Lussac. In the latter equation, $pv = Rt$, when $p =$ pressure, $v =$ volume, $R =$ the gas constant, and $t =$ the temperature. The suggestion is that income density is like pressure; area is the quite reasonable counterpart of volume; the total population is like the gas constant in that this constant is proportional to the molecular population; and per capita income is temperature-like. At the macroscopic level, therefore, the analogy is reasonable.

The contact of the physicist with gases at first was at their macroscopic level, because only macroscopic measurements could be made. Boyle, for example, was quite unaware of the actual composition of the samples of gases which he studied. However, subsequent inference and later verification of the microscopic nature and behaviour of the gases resulted in a kinetic theory. In this theory the microscopic or molecular nature of the gas is stressed and the macroscopic phenomena become merely statistical averages. Temperature and pressures are explained in terms of the intimate association of kinetic energy with the molecules. Temperature is proportional to the average energy per molecule with pressure the energy per unit volume.

This leads us to the recognition of income as energy-like; i.e. a possible form of social energy. However, income cannot properly be considered as a social energy since it is a flow and has a time dimension. A preferable measure of social energy is capital, publicly or privately owned and broadly defined to include human intelligence, which, when exerted through time, produces income.

For the purposes of the present study, it is therefore important to investigate the geographical distribution of wealth; i.e. stocks of capital. Despite acute data problems such an investigation has been made and will be described subsequently. Income is then to be analyzed at a higher level of abstraction, i.e., the "action" level.

A general law of energy change (whether social or physical energy) is that it may be viewed as a product of an intension times the change in extension. Thus, per capita income wealth in dollars per person has temperature-like dimensions and is the intension of the heat-like total wealth with number of persons as the appropriate extension. On the other side of the equation, wealth density in dollars per square mile is seen to be like pressure. Complementing this intension is the extensive factor, that *sine qua non* of geography, area. For the time being, let us continue to consider income as energy-like and defer the discussion of wealth for the moment.

The gas laws as stated above refer to a perfect gas in equilibrium in which there is uniformity of temperature and of pressure throughout. Our examination of the income density and per capita income surfaces, of course, discloses gradients corresponding to the geographical variation of the income potential surface. In the early Eddington model of a star, temperature is proportional to the gravitional potential, but discontinuities in the temperature surface are not considered. Pressure continuity with temperature discontinuities, however, is a feature of the well-known and widely published daily weather maps.

The boundary which separates masses of air unlike temperature, but is associated with no discontinuity of the pressure surface across it, is designated a front. For the 1967 distribution of income in the United States the boundaries separating the T' from the T'' regions may be regarded as "Income Fronts" and the regions as "normal", "hot", and "cold", respectively. The lines of discontinuity on the per capita income surface establish these fronts. It is again to be observed that, to a first approximation, no such interruptions to the income density surface occur.

Certain details of the agreements and disagreements involved in the analogy between the income fronts and the air mass fronts are included below. The phenomenon now labelled income front was discovered independently, with no *a priori* definition, and its importance here rests solely upon the role it has in the analysis of the geographical distribution of economic and social phenomena, however.

Certain objections will surely be raised about the averaging of data by states in this study. Preliminary examination of available county data, however, suggests that only two changes in the above presentation would be necessary: (1) a refinement of the location of the major fronts, which in the first part of the study were of necessity drawn on a state boundary segment basis; and (2) the determination of certain minor fronts. These findings are to be presented in detail at a later date.

When the microgeographic distribution of population and income is regarded in a very detailed urban-rural distribution, minor local peaks of potential agree with local high densities. The higher per capita incomes of urban over rural areas thus follow the general rule already established in this study. All of the surfaces described remain the same in their broad configurations. Detail is added but geographical coincidence remains. And the general income fronts are as before, albeit with greater locational precision.

Not only are the major regional fronts observable, but within cities themselves similar "Slum Fronts" can be discovered. Income densities (and the necessarily closely-allied areal rents and land values) can be seen to behave as in an areal continuum. Per capita income "regions" in the cities do emerge with associated discontinuities. Slums can thus be demonstrated to produce the required equilibrium amount of dollars per unit area, but only by means of the characteristic high population densities, i.e., the extreme crowding of great numbers of low per capita income persons into smaller areas.

(e) Singular points, lines and areas of the
1967 income potential surface

Given here in fig. 5.6 is a more detailed map of Income Potentials for the United States based on data effective as of December 31, 1967. Computations were made for values at each of 3070 control points, these representing data grouped for county or county-like areas. Isopotential lines were then interpolated among the values computed for control points. These isopotential lines on this "socio-economic terrain" may be interpreted as analogous to contour lines on the physiographic terrain.

Contour lines result from the intersection of surfaces, conventionally the intersection of a variable surface with specified constant-valued "level" surfaces. Any one line thus connects points of equal value on the variable surface and is distinguished by a numeral on the variable surface showing the level surface to which it belongs (fig. 5.7). The surface represented by figs. 5.7-5.12 is, of course, a hypothetical one used for illustrative purposes.

The contour at a peak is a point. A peak is a local maximum of elevation. Everywhere in the immediate neighbourhood on the surface elevation values are lower.

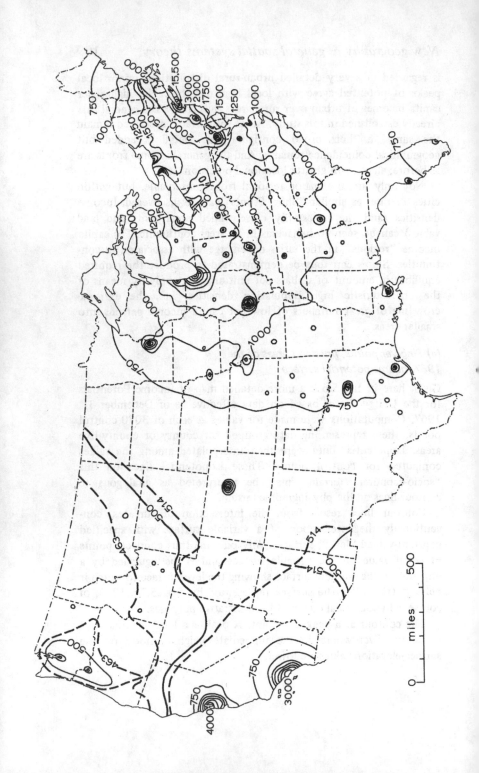

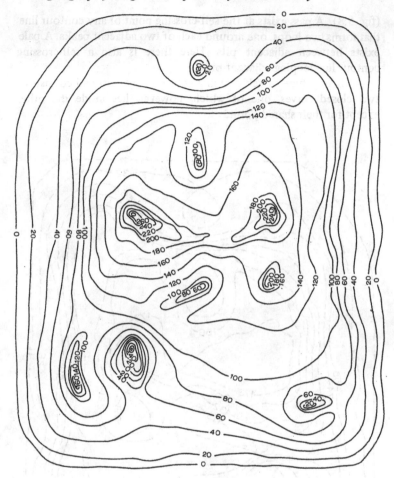

Figure 5.7

The contour at a pit is a point. A pit is a local minimum of elevation. Everywhere in the immediate neighbourhood on the surface elevation values are higher.

Peaks and pits are singular points constituting the category, "absolute extremum points". The other kind of singular points are "mixed extrema points" and are designated passes and pales (fig. 5.8). A pass exists at the self-crossing point of any contour line that forms two loops, one around each of two adjacent peaks. A pale exists between adjacent pits. Here there is also a self-crossing contour line looping adjacent pits.

An inloop type of self-crossing contour may exist and consists of two closed curves, one of which, however, lies inside the other except for their shared point.

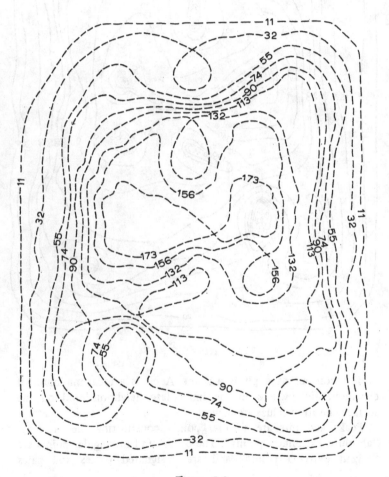

Figure 5.8

Through every point on the surface there is not only a contour line, but also a slope line indicating the direction of steepest gradient and, of necessity, intersecting the contour lines at right angles. In general, any one slope line leads to some peak in its "uphill" direction and to some pit in its "downhill" direction. If, however, a particular slope line is found to run from a pit to either a pass or pale, then that line, when continued through that pass or pale, will, of necessity, run only to another pit (or in rare cases, the same pit). Similarly, any slope line found linking a peak to a pass or pale, must then, when extended through the pass or pale, continue only to another peak (or in rare cases, the same peak).

Slope lines linking peaks via passes or pales are designated "ridge lines" (fig. 5.10). Slope lines linking pits via passes or pales are designated "course lines" (fig. 5.9). One profile shows the pass at the lowest values between two peaks; a profile through the same point taken at right angles to the first shows the value at the pass to be higher than any other along that cross section. The pale occurs at the high point between pits, but it is a low point along the line at right angles to the line joining the pits. Passes and pales are thus "mixed extrema points" and, collectively, are called knots.

A given surface may be completely divided into the hills that constitute it. Shown in fig. 5.9 are individual hills, each with its own peak, with course lines defining the boundaries and with the pits, passes, and pales of the surface on these boundaries. An exceptional case occurs within hill number 4 which is shown on other maps to have one of its set of ridge lines with an intervening knot beginning and ending at its peak. All hills are bounded by course lines, but not all course lines bound hills.

Independently, the surface may be divided into dales, having internal pits and with peaks, passes, and pales, lying on the ridge line boundaries (fig. 5.10). Although not shown here, it may happen that one set of course lines with an intervening knot begins and ends at the same pit. In this case there will be a ridge line not constituting a part of the boundary of some dale.

Every point on the surface, save those on course lines or ridge lines, lies *simultaneously* within *both* some hill *and* some dale.

Territories formed by boundaries composed of both course lines and ridge lines are of interest (fig. 5.11). In general, a territory has two course line segments and two ridge line segments constituting its

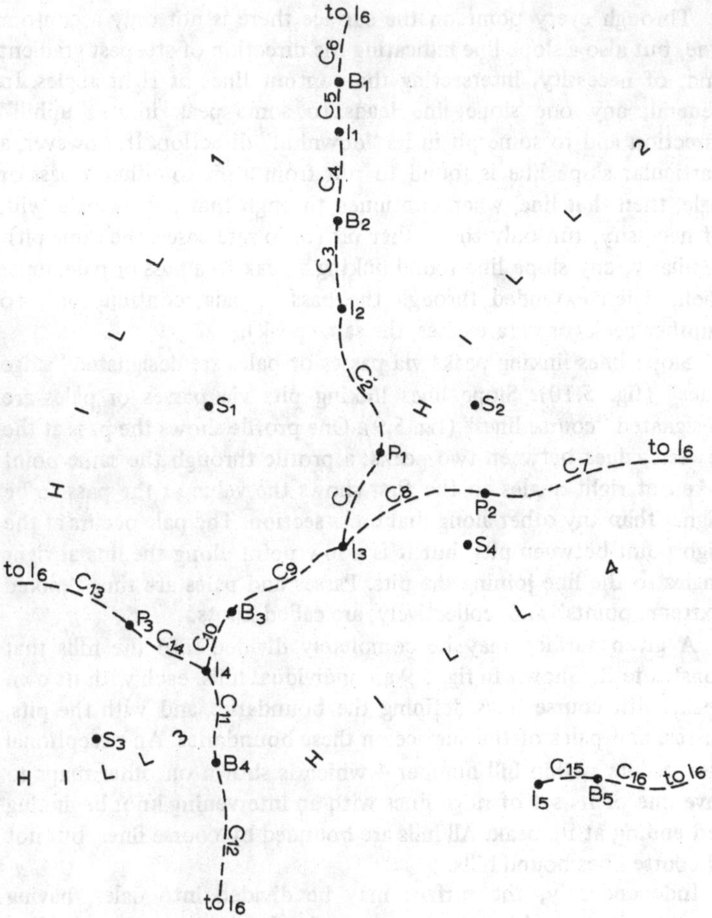

Figure 5.9

total boundary. Thus, all singular points are found on boundaries
and not in the interiors of territories. The territories, in general,
contain points that have a smaller extension than hills or dales taken
separately. Exceptional cases exist, as implied above, when a single
dale or a single hill constitutes a territory. In our sample surface
illustrated here, dale 5 is seen to be exactly coincident with territory
XII (fig. 5.12 is a composite).

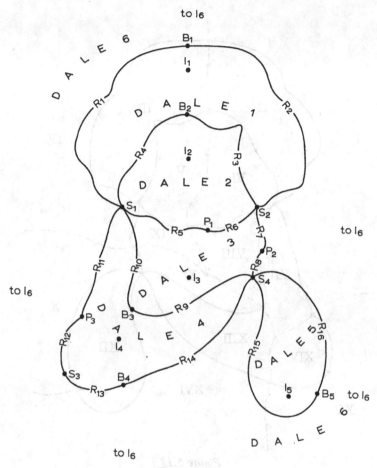

Figure 5.10

Within any completely closed contour line on a continuous surface the number of peaks, S, is always one more than the number of passes, P, so that $S = P + 1$. The same rule applies to the number of pits, I, and the number of pales, B, so that $I = B + 1$.

If, in the singular cases of passes and pales, we count each of these as single, double, or n ... ple, depending on whether two, three or n+1 areas of elevation or depression meet at a pass or pale,

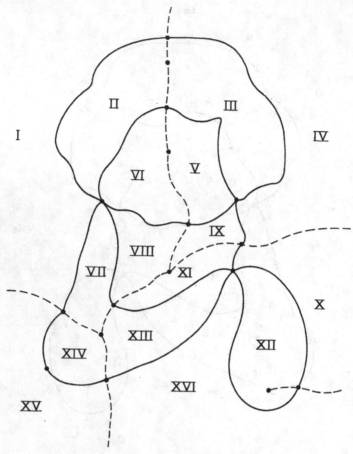

Figure 5.11

respectively, then the above counts can be taken as before, giving each singular points its proper number.

With reference to the area enclosed by any one contour line, the following obtains: $(S + I) - (P + B) = 1$. However on a closed surface like a spherical distribution, $(S + I) - (P + B) = 2$. For, on a sphere, a closed curve bounds two areas. For a general topological consideration, let V be the number of all singular points on the closed surface ($V = S + I + P + B$); let E be the number of lines ($E = R + C$); and let F be the number of separate territories. Then, $F = E - V + 2$.

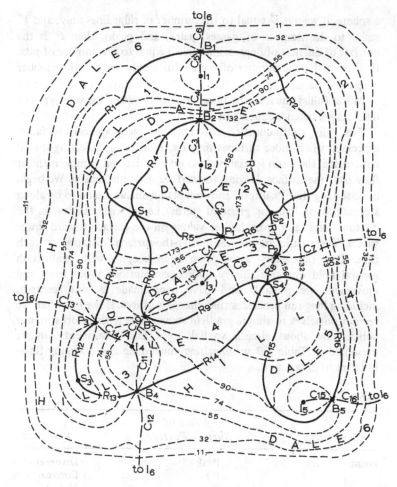

Figure 5.12

(Again, within any one closed contour line, $F = E - V + 1$). This general topological relationship among points, lines, and areas was first established by Euler in network analysis (the "Königsberg bridge problem") and was explained in terms of any polyhedron where the number of faces plus the number of vertices minus the number of edges equals two.

Further consideration of the above shows that, for example, over a sphere, it we put E' equal to the number of ridge lines only, and V' equal to the number of passes, pales, and peaks, then F' is the number of districts of depression (dales) equal to the number of pits. If E'' specifies the number of course lines only and V'' the number of passes, pales, and pits, then F''' is the number of districts of elevation (hills) equal to the number of peaks when the two types of districts are taken independently.

For all continuously differentiable geographical surfaces, especially when those surfaces may be regarded as indicating spatial structuring, short-run spatial process may be defined as movement or flow over the surface, leaving the surface itself unchanged. We begin by assuming that "natural" movements on surfaces tend to be along steepest slope lines or gradient paths, i.e. at right angles to the contour lines, and from higher values on the surface toward lower values. These paths are minimum over-the-surface "distance" in each case. This elementary assumption is in keeping with analysis in general field quantity theory. Other assumptions about form and movement are possible and have been made in our research, including long-run processes that change surfaces. However, when we restrict ourselves to simple gradient movements in the short-run, our conclusions about simple spatial form and movement can be presented in Table 5.3 summarising converging and diverging flows in terms of dimensions.

Table 5.3

Dimension	Name of Surface Feature	Vergency
Point	Peak	Divergence
	Pit	Convergence
	Pass	Mixed
	Pale	Mixed
Line	Course	Convergence
	Ridge	Divergence
Area	Hill	Divergence
	Dale	Convergence
	Territory	Mixed

Applications within a "general spatial systems" theory have been made to physical land forms and fluvial systems, populations and economic flows, central place hierarchies, human migrations, meteorological data, oceanographic data, and various other spatial patterns.

The data portrayed in fig. 5.6 reveal that among the 3070 control points 184 qualify as peaks, 183 as passes, 51 as pits and 50 as pales. Of course the maps does not capture them all because of the contour interval chosen and, in addition, the method of population grouping (i.e., 3070 control points) causes minor features to be obscured. Thus the minor orders of features are not portrayed.

The 3070 control point values were analyzed and reported upon by the research group (Kingsbury 1971). Summarized information is as follows.

Let it be assumed that the least complex surface would be one on which no singular points were contained. Let this condition also be defined as the most probable and the least ordered condition.

Figure 5.13 shows a graph relating cumulative numbers of singular points to cumulative total number of control points, including singular and non singular when the accumulation is done in order from highest to lowest values of income potential. Specifically, $S = .55C^{.83}$ with complexity thus greater at higher values of income potential. These data provide the basis for computing the negentropy or information content of the surface (See Kingsbury 1971).

Figure 5.6 implies the location of the singular points of the income potential surface. A study of the detailed data shows that the pass with the highest income potential value is that linking the New York peak (maximum maximorum) to the Philadelphia peak.

The New York-Philadelphia combination and its accrued area is connected by a pass in Cecil County, Maryland to the accrued area of the combination including the Baltimore and Washington, D.C. peaks with their pass thereby creating a combination with four peaks and associated area. Such a procedure as this contains the means for establishing an ordering of the features of the income potential surface.

The next three passes connected to this New York-Philadelphia and Baltimore-Washington combination all attach single peaks to it. As an experiment in ordering we might attempt the following.

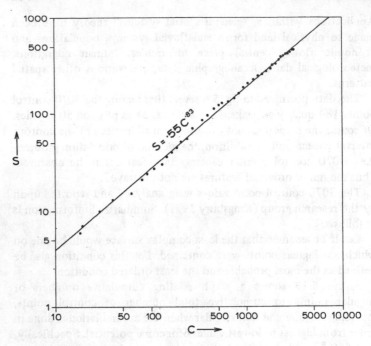

Figure 5.13

Let each pass be assigned an order equal to the lower of the orders of the two peaks or combinations connected. Thus, the New York-Philadelphia pass is of the first order, the pass in Cecil Co., Md. is second order, and the three passes connecting single peaks to the third order combination are all first order.

The fourth and highest order combination is created when New York-Philadelphia plus Baltimore-Washington is connected to Detroit-Toledo plus Cleveland-Pittsburg through the third order pass in Centre Co., Penn. One other third order pass exists, connecting the fourth-order "New York Massif" to the third order combination of Chicago-Milwaukee plus Indianapolis-Madison Co., Indiana. Another important pass is the second order pass connecting San Francisco-Los Angeles to the New York Massif. There are a total of fifteen second order passes on the potential surface, and 166 first order passes.

For each of the 183 passes, there is a self-crossing or "figure-eight" contour of equi-potential which intersects itself at the pass. The two loops of the figure-eight enclose the two peaks or combinations which are connected. If the pass connects two peaks, or connects a single peak to a larger combination, the area enclosed in a contour loop surrounding a given peak is said to be "associated" with that peak. Consider a second order pass connecting two second order combinations, or connecting a second order combination to a larger combination. All the area enclosed within a loop surrounding a second order combination, but lying outside the loops surrounding single peaks, is said to be associated with the second order combination. The same approach is applied in determining third order associated areas. The majority of the area in the United States is thus considered to be associated directly with the fourth order New York dominated Massif. In this way all the area in the country is associated with some peak or combination whether in a direct or intermediated way. Letting a county control point value be representative of the county area average value, we can proceed as follows.

A county associated with a single peak has an income potential which is less than that of the peak county but greater than that of the pass connecting the peak to another peak or combination of peaks, and shares a boundary either with the peak county or with another associated county of higher income potential.

A county associated with an nth order combination of peaks has an income potential which is less than that of the pass connecting the nth order combination to another combination of at least n orders, and shares a boundary with a peak, pass, or associated county of higher income potential belonging to the nth order combination.

Such a procedure is not altogether satisfactory, however, as the values of income, population, income potential integrated through area ($\int U dA$) etc. associated with the combinations should also enter into the definition of the order of passes, i.e. total social energy flow conditions or processes in addition to the structures. Moreover, such a procedure associates lower order with higher importance and this has always been a source of difficulty in defining the ultimately defensible system of hierarchical ordering.

Population as a thermodynamic aggregate

We call attention again to the expression $DA = PT$ and the similarity to the equation of state for the perfect gas, $pv = Rt$. This time, however, we recast the thermodynamic equation of state, following Fein (1970) as $pv = nRt$, where p, v, and t are state variables, n is a measure of molecular quantity and R is a constant determined by the units of the state variables. Molecular density, G, is then, $G = n/v = p/Rt$. (G is substituted here for Fein's symbol q as q has already been assigned another role, as a dimensionless constant in macrogeographical research. Other convenient notational changes are also made below).

Fein notes that if the aggregates of the system are acted upon by a central force which is attenuated inversely as the square of distance, or from a spatial point of focus, the following differential equation will represent the equilibrium density distribution.

$dG = (1/Rt)dp = (1/Rt)(-Kr^2)G(r)dr$ where K is a constant of the force field and r is, as before, distance. It is important to note that a force varying inversely as the square of the distance is *entirely* consistent with a definition of potential involving the inverse of distance. The derivative of the potential at any point with respect to distance yields a measure including the inverse of distance squared. This is the gradient of the potential (scalar field) and is the intension of the force (vector field).

Let us now recast the Fein expression recognizing our earlier identification of the macrogeographic distributions with the equation of state and the identification of I with the energy of the system.

$$dG = (1/RT)dD = (1/RT)(-K/r^2)G(r)dr$$

Integrating this equation and recognizing a boundary condition $G = G_c(r_c)$ where r_c is the distance from some central point beyond which no "molecules" are found allows application to city structures of population density. We find that $G = G_c \exp(B(r_c - r)/rr_c)$ with $B = K/RT$.

For cases where $r \approx r_c$ for all of the area of interest, the above equation simplified to $G = G_c \exp(B'(r_c - r)$ with $B' = KR'/T$.

Thus, the drop off of population density can be seen to vary inversely with per capita income T, for the identification of K/T with

B links the thermodynamic equation of state of a population precisely with the earlier empirically determined law of exponential decrease outward for urban population densities found independently by Clark (1951) and Stewart (1953) and elaborated upon by Stewart and Warntz, for B above is the coefficient found. The levelling out of population distribution in cities as inversely related to per capita income has been checked for five U.S. cities by Fein following Clark's data. B decreased by a factor of about one-half between 1910 and 1940. For the same period, T (in constant 1929 dollars) rose from 400 to 720. Additional graphical evidence for Chicago, 1860 to 1950 is given by Fein and is most persuasive. Of course, the equations should be tested separately against each of the three regions bounded by the income fronts to see if earlier significant differences in constants are being modified toward overall similarity in urban spatial structures (and processes) accompanied by the dissipation of the fronts.

The total energy change (in the form of heat) of a thermodynamic system may be given as

$$dI = CdT + DdA$$

where C is a constant of the system. (Again we have made the isomorphic substitutions identifying income as energy-like with per capita income to be equated to "social" temperature, income density to "pressure" and with area as the geographical extent of the system. Again we caution that capital or wealth is the appropriate isomorph of (heat) energy and that income is action.

It can be noted that for an adiabatic change in the thermodynamic variables, in which there is no exchange of heat between the system and its surroundings, $dI = 0$, and the above equation will yield a solution of the form,

$$T = sD^m$$

where s and m are both positive constants for a given system.

We have already established the empirical relationships

$$D = kU^w \text{ and } T = zU^x$$

where D is income density, U is income potential, and k, w, z and z fitted positive constants. Substituting for U in the second equation the solution obtained from the first gives,

$$T = \text{constant } (D^{x/w}).$$

This being so this equation is seen to be equivalent to

$$T = sD^m$$

and may be written, upon substitution of $D = TG$, as

$$T = \text{constant } G^{x/(w-x)}$$

where G is again the population density.

Appropriate computation again reveals the regionalization of the constant and the existence of the income fronts, particularly the relative isolation of the traditional "south". This is, of course, no new information as it follows mathematically from the equations established earlier from the empirical data. In the immediately above form, however, we gain additional insights into the nature of the American economy varying through time and of the spatial aspects of the system. Particularly, the fact that the exponent of G with which T varies is positive, allows us to assert that the relationships among T, G, and D are adiabatic within "regions". (Yet another definition of economic regions!) We follow Fein's suggestion that we may take as a measure of an area's "insulation" the fact that certain demographic variables can be fitted by an adiabatic equation. If there is an external reservoir across which energy flows during the thermodynamic change of state, the process will not be adiabatic. The development of new areas into which capital or wealth is poured may be an example of this. Observation of adiabatic processes implies, then, a degree of economic self-sufficiency for the area. We propose to study subsequently additional aspects of capital and population migration flows and the relationships of adiabatic and non adiabatic processes to social order and entropy.

Conclusions

Lack of space here prevents as well the exploration of other topics from General Systems Theory found naturally to be applicable through previously independent research in macrogeography and its original parent, social physics. These would suggest further that the discipline of geography be regarded as General Spatial Systems Theory. Additional matters of importance in which contact has been made with General Systems Theory relate to Dimensional Analysis and the determination of the constants (dimensionless and dimensional) of a system, information and entropy (not to be regarded as either similar or inverses), isometric and allometric growth, homeostasis and "feedback" controls, open and closed conditions, and so on. In each of these, empirical findings reveal their significance. Moreover the findings come from all parts of the traditional "fields" of geography, e.g. social, physical, economic, etc. Together they constitute evidence that geography has much to benefit from thorough application of existing concepts from general systems theory and also that G.S.T. itself stands to benefit from the work of geographers. The integrity of geography as a discipline will be enhanced if it is realized that within the G.S.T. framework geography finds its place as responsible for and sensitive to General Spatial Systems Theory when the space involved is geographical. Such an organization of the discipline would make possible communication and understanding across the specialty and subspecialty divisions within the discipline and equally among geography and other disciplines. This is admittedly an optimistic view and perhaps naïve as well in that the limitations are not identified and discussed. For a balanced view the reader is referred to an excellent paper by Anderson (1969) who carefully reviews and criticizes relevant aspects of G.S.T. with regard to urban geography and who supports Davis (1966) that at this early stage the "significance of the interdisciplinary movement to geography rests as much on the mental attitude it engenders as on concrete results achieved to date."

References

ACKERMAN, E.A. (1965) *The Science of Geography*; (Washington, D.C.).

ANDERSON, J. (1969) On general systems theory and the concept of entropy in urban geography; *London School of Economics Graduate Geography Department Discussion Paper No. 31*, 17p.

BERRY, B.J.L. (1964) Cities as systems within systems of cities; *Papers and Proceedings of the Regional Science Association* 13, 147-63.

BERTALANFFY, L. von (1968) *General Systems Theory*; (New York), 289p.

CHORLEY, R.J. (1962) Geomorphology and general systems theory, *U.S. Geological Survey, Professional Paper* 500-B, 10p.

CLARK, C. (1951) Urban population densities; *Journal of the Royal Statistical Society* 114 (4), 490-6.

DAVIS, W.K.D. (1966) Theory, science and geography; *Tijdschrift voor Econ. En. Soc. Geografie* 57, 125-30.

FEIN, E. (1970) Demography and thermodynamics; *American Journal of Physics* 38, 1373-9.

KINGSBURY, D. (1971) A Description of the 1967-68 United States income potential surface; *Harvard Papers in Theoretical Geography, No. IV, Geography of Income Series*, 13p.

STEWART, J.Q. (1953) Urban population densities; *Geographical Review* 43, 575.

STEWART, J.Q. and WARNTZ, W. (1958a) Macrogeography and social science; *Geographical Review* 48, 167-84.

STEWART, J.Q. and WARNTZ. W. (1958b) Physics of population (Regional Science Research Institute, Philadelphia), 117p.

WARNTZ, W. and WOLDENBERG, M.J. (1967) Concepts and applications – Spatial order; *Harvard Papers in Theoretical Geography, No. 1, Geography and the Properties of Surfaces Series*, 189p.

WOLDENBERG, M.J. and BERRY, B.J.L. (1967) Rivers and central places: Analogous Systems?; *Journal of Regional Science* 7, 129-39.

6 Some questions about spatial distributions*

MICHAEL F. DACEY

This paper poses questions about properties of spatial distributions. Because the objective of these questions is to stimulate study of the fundamental properties that distinguish and discriminate spatial distributions, it does not attempt to provide answers.

It may appear that this paper also provides a logically coherent structure for examination of spatial distributions. If so, it is because formal terminology and symbolic notation are confused with rigor and precision of expression. In fact, there are numerous inadequacies, such as use of functions, relations and concepts that are poorly defined. It is anticipated, but in no sense demonstrated, that it will be possible subsequently to construct suitable definitions while retaining a logically consistent statement.

Definitions

The region R is located on the two-dimensional euclidean space E. Though not necessary, it is presumed in the following that R is a bounded, connected region with finite area $m(R)$. Other special properties of R are indicated as the need arises.

The collection $\Omega_n = \{\omega_1, \ldots, \omega_n\}$ contains n distinct objects. The attributes that distinguish members of the collection are inconsequential, but a critical requirement is that the objects are punctiform in the sense that their locations may be specified at any required level of definiteness.

The collection $F(R,\Omega_n)$ of one or more location rules specifies the manner of locating within R the objects belonging to Ω_n. If there is no risk of confusion, F is written instead of $F(R,\Omega_n)$.

* The support of the Office of Naval Research, Contract N00014-67-A-0356-0009, is gratefully acknowledged.

It may appear that the location rule $F(R, \Omega_n)$ is an appropriate topic of study. If study is restricted to F, many interesting problems of geographic analysis are excluded from consideration. A frequent starting point for many geographic investigations is a list or map giving the locations of all objects, of a specified kind, that occur in a geographic region and the inferential problem is to discover a location rule that accounts, in some specified manner, for the observed locations. It is difficult to use the notation $F(R, \Omega_n)$ and express that F is unknown. The notation $F(R, \Omega_n)$ is also cumbersome when there is need to identify a spatial distribution formed by the locations of the objects in Ω_n that are located on a subregion of R. For these and similar reasons, it is advantageous to consider the triple defined by a region R, a collection of objects Ω_n and a location rule F.

Definition 1. Suppose R is a region, Ω_n is a collection of $n > 0$ objects and F is a location rule. If $R \supset S$ and $m \leqslant n$, then the triple $(S; \Omega_m; F(R, \Omega_n)) = (S; \Omega_m; R, \Omega_n)$ is called a spatial (point) distribution. For brevity, the triple representing a spatial distribution is denoted by D. If $R = S$ and $m = n$, the notation simplifies to $(R; \Omega_n; F)$.

Definition 2. If the location rule F contains a chance or probabilistic component, D is called a stochastic spatial distribution; otherwise it is called a deterministic spatial distribution.

Example 1. Let R be any region and let Ω_n be any finite collection of objects. Suppose $F(R, \Omega_n)$ is defined in the following way: each object in Ω_n is uniformly and independently located on R. Then $(R; \Omega_n; F)$ is a stochastic spatial distribution and is called a random distribution (of n objects on R).

Example. Suppose $(R; \Omega_n; F)$ is a random distribution and $A \subset R$. Then $(A; \cdot; R, \Omega_n)$ is a random distribution on A but the number of objects on A is not fixed, but instead is a random variable.

For arbitrarily defined components, it may be necessary to verify the existence of D. Such existence problems are avoided here by restricting examples to triples for which D is known to exist.

The stochastic spatial distribution D is a mathematical construct. While it can be described in terms of measurable properties, it is not possible to display it as a map or other physical representation. Because it is sometimes convenient to be able to interpret a stochastic spatial distribution as a mappable distribution, a map representation is defined by a realization of D.

Definition 3. A map representation $m(D)$ is a cartographic display of a realization of a stochastic spatial distribution D.

It is recognized that $m(D)$ is only a sketch that combines cartographic conventions with a synthetic simulation that approximates a realization of D.

It is also convenient to be able to specify a map representation of an observed distribution of objects.

Definition 4. Suppose R is a region of the earth's surface and W_n is a collection of n objects that are located on R. If these objects are represented on a map M by "point" symbols designating, at a given map scale, arbitrarily definite locations, and R is represented on M as a portion of euclidean space, then $M(R; W_n)$ is called a (spatial) map distribution of an empirical distribution of (punctiform) objects. The notation $M(R; W_n; F)$ is used to indicate that the locations of objects in W_n are the consequence of a location rule F.

It is an untestable assumption that M is a realization of a stochastic spatial distribution. However, if it is assumed that M was generated by a location rule, however complex, that contains a chance component, then it is legitimate to seek to identify a stochastic location rule F such that M may be treated as a realization $m(D)$ of a spatial distribution D.

Example. The observed map distribution M(Iowa; 99 largest places, 1950; F) has been examined by Dacey (1964a). The available evidence does not contradict the assertion that this map distribution may be treated as a realization of (region shaped like Iowa; Ω_{99}; F), where F represents the location rule specified by Dacey's county seat model.

The term "map distribution" and "stochastic spatial distribution" are used to discriminate observed and theoretical spatial distributions. The term "spatial distribution" is used to refer to both observed and theoretical spatial distributions or when the distinction between observed and theoretical is either obvious or not critical.

Objectives

This paper was initially conceived as a more complete statement of the preceding structure, followed by a review and evaluation of stochastic models available for formulation and specification of spatial distributions. As such, the intention was to concentrate on formulations of stochastic location rules F and the identification of properties of the distribution D resulting from objects located in accordance with these rules. There currently are available many stochastic models that are, or could be, used in the study of spatial distributions. In addition to models that have been especially constructed to account for components of location processes, many other models are readily interpreted as statements of locations in a spatial situation. Frequently, these are two- and three-dimensional extensions of models initially formulated in a temporal context. Because many of the models have not been explicitly adapted to the study of spatial distributions, current understanding of stochastic spatial distributions would be augmented by a unified and consistent explication of the spatial implications of a wide variety of stochastic models. Such a study ideally would use a common body of concepts, terminology and notation to identify for each spatial distribution both its underlying location rule $F(R; \Omega_n)$ and its properties, such as occupancy and spacing measures, that are measurable and comparable with observable properties of map distributions. The rationale for this type of paper is that a foundation for more advanced and penetrating studies of stochastic location theory would be provided by a systematic compilation, even though incomplete and highly eclectic, that succeeds in integrating diverse models into some sort of coherence.

The utility of this type of compilation is less clear when considered with respect to its potential contributions to the geographic study of map distributions. To serve these needs a compilation must be selective and emphasize the properties of location rules and stochastic spatial distributions that are germane to the description and analysis of map distributions. The primary focus of this empirical analysis is the comparison of observed spatial distributions with theoretical spatial distributions and with other observed spatial distributions. A typical investigation is the

evaluation of available empirical evidence to ascertain if it supports
the hypothesis that an observed map distribution may be treated as a
realization of a stochastic spatial distribution in the sense that the
map distribution has properties similar to those of a specified,
theoretical spatial distribution. Or, properties of two or more map
distributions may be compared in order to evaluate the hypothesis
that they are realizations of a common, but unknown, spatial
distribution. A more theoretical type of study concerns the ability
to distinguish map distributions that are properly treated as
realizations of different spatial distributions. The common theme of
these and similar geographic investigations is the description and
classification of spatial distributions.

It is ironic that the literature of geography, which is frequently
defined as the study of "areal differentiation" or "areal
distributions," does not provide a single, well-developed conceptual
and methodological framework that organizes and structures the
description, classification and analysis of spatial distributions. For
example, the basic methodological statements by Hartshorne (1939),
Bunge (1965) and Harvey (1969) almost totally ignore the study of
map distributions and its role in the construction and testing of
geographic theory. One consequence of the lack of an appropriate
methodology is that it is not clear what properties and relations need
to be emphasized in a compilation of location rules and stochastic
spatial distributions that is intended to promote and facilitate the
geographic study of map distributions. Because criteria of pertinence
are not available, the risk is that such a compilation will stress trivial
and irrelevant properties and relations, while neglecting the
distinctive and distinguishing and, hence, will not provide the types
of information that are relevant to geographic studies. This is one of
the primary reasons that this paper did not become a review and
summary of stochastic models adaptable to the study of spatial
distributions. Instead, this study has been redirected to the
seemingly more productive task of examining some of the basic
concepts and assumptions that underlie the comparison of map
distributions.

The strategy for this investigation is, possibly, unorthodox. While
there is the customary review of the existing literature, this literature
is scant and not particularly informative, so that this study cannot

be related to a geographic tradition and formative literature that provides both guidance and constraints. As a consequence, the focus of this study is a largely unexplored topic for which there does not even exist an appropriate terminology and carefully formulated body of definitions. So, it is not possible to relate this examination of map distributions to a conventional, well established structure. While an alternative option is to organize the basic concepts with respect to an arbitrarily defined structure, this approach presumes ability to select properties and relations that explicate a concept that is itself unformulated. It is difficult to have confidence in a conceptual and methodological framework that is largely speculative and anticipatory. Instead, there is a prior need to have a strong intuitive grasp of the fundamental properties and classes of spatial distributions. In a sense, the implication is that, at this time, it is not productive to construct a methodological framework for spatial distributions. Possibly, though, it is productive to supply questions that stimulate inquiries into the basic nature of spatial distributions. The approach adopted in this study is to pose questions, but not attempt to answer them.

Methodological frameworks: the continuum and independent components hypotheses

Currently available formulations of the concept of spatial distribution make extensive use of two different sets of diagrams to convey its salient attributes. One set suggests that spatial distributions form a continuum, while the other set suggests that spatial distributions are composed of several independent components.

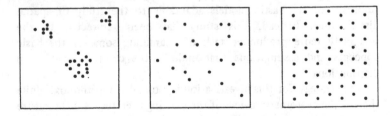

Figure 6.1 The diagrams illustrate, from left to right, clustered, random and regular spatial distributions.

Continuum hypothesis

One approach is to display three diagrams (fig. 6.1) which are labelled *clustered distribution, random distribution* and *regular distribution*, or similar terms. The random distribution in this display corresponds to a realization of the random spatial distribution identified in Example 1. The basic assumption that motivates the use of these diagrams is that the random distribution is a midpoint on a continuum of spatial distributions that vary from completely clustered to totally regular.

The following applications of the continuum hypothesis are highly simplified, but they illustrate the basic steps involved in the classification of spatial distributions. Though the examples are formulated in terms of map distributions, essentially similar procedures are used to classify theoretical distributions.

Example 2. For an observed map distribution $M(R; W_n)$, let $x(A)$ represent the number of objects located on a subregion $A \subset R$. For disjoint regions $A_1, \ldots, A_m$, which may or may not constitute a sample, put

$$\bar{x} = m^{-1} \sum_{i=1}^{m} x(A_i),$$

$$s^2 = m^{-1} \sum_{i=1}^{m} [x(A_i) - \bar{x}]^2,$$

and $\epsilon_m > 0$ is a parameter that is estimated with use of some statistical theory. The quantity $s^2/\bar{x}$ is used to classify M. A crude, but commonly used, classification uses the following decision rules.

(a)   If $1 - \epsilon_m \leqslant s^2/\bar{x} \leqslant 1 + \epsilon_m$, then M is random.

(b)   If $s^2/\bar{x} > 1 + \epsilon_m$, then M is more regular than random.

(c)   If $s^2/\bar{x} < 1 - \epsilon_m$, then M is more clustered than random.

Example 3. For the map distribution $M(R; W_n)$, let y_i represent the distance from w_i to the nearest other object in W_n. For distinct objects $w_1, \ldots, w_m$, $m \leqslant n$, which may or may not constitute a sample, put $\bar{y} = m^{-1}(y_1 + \ldots + y_m)$ and $c^2 = n/m(R)$, and $\epsilon_m > 0$ is a parameter that is estimated with use of some statistical theory. Then $\bar{y}$ is used to classify M, and the following is a typical classification.

(a)   If $\frac{1}{2}c - \epsilon_m \leqslant \bar{y} \leqslant \frac{1}{2}c + \epsilon_m$, then M is random.

(b)   If $\bar{y} > \frac{1}{2}c + \epsilon_m$, then M is more regular than random.

(c)   If $\bar{y} < \frac{1}{2}c - \epsilon_m$, then M is more clustered than random.

Though these examples may be improved by a more sophisticated mathematical theory that may take into account the shape of R and the size of n, they suffice to explicate the essential features of the continuum hypothesis. This formulation presumes that spatial distributions form a continuum from completely clustered to totally regular and that the position of a spatial distribution on the continuum is identified in terms of cell counts or nearest neighbor distances. The theoretical and mathematical studies engendered by this approach to spatial distributions are largely concentrated on three topics: the selection of size of the cells A_i used in the cell count method; the derivation of properties of cell counts and nearest neighbor distances for theoretical spatial distributions; and the statistical theory underlying the specification of ϵ_m.

One obvious defect in this formulation of spatial distributions is that, on the basis of cell counts, it classifies the random distribution of Example 1 as a distribution more clustered than random. More importantly, it is difficult to concur with the hypothesis that positions of spatial distributions on a linear continuum either generates a useful classification scheme or confirms intuitive notions of degree of similarity. At the least, there is need for substantial supporting evidence, which thus far has not been provided.

Independent components hypothesis

The only other widely used formulation evolves from the premise that spatial distributions are composed of several components which may be varied independently of each other. The most cogent statement of this formulation appears in a high-school level textbook prepared by Thomas (1965). The model is expressed in terms of properties of spatial distributions of facts which are of significance to geographers. These facts are called *geographic facts* and are defined by Thomas as "facts which indicate the quantity or quality of a specified phenomenon which occupies a specified place at a specified time" (p. 14). The independent components that characterize the spatial distribution of a collection of facts are called *pattern, density* and *dispersion* and are defined (p. 87) in the following ways.

The *pattern* of a spatial distribution is the areal or geometric arrangement of the geographic facts within a study area *without regard* to the size of the study area.

The *density* of a spatial distribution is the overall frequency of occurrence of a phenomenon within a study area *relative* to the size of the study area.

The *dispersion* of a spatial distribution is the extent of the spread of the geographic facts within a study area *relative* to the size of the study area.

The "facts" are usually interpreted, as Thomas does, as the presence of objects with arbitrarily definite locations. In this case, Thomas' use of "spatial distribution" is compatible with the terminology adopted in this study.

Diagrams (fig. 6.2) are used to illustrate these concepts and to demonstrate that they may be independently varied.

A similar conceptual framework is identified in a number of subsequent studies, such as Hudson and Fowler (1966) and Rogers (1969), though the structure is usually modified by deleting density from the list of independent components and sometimes using alternative terms for pattern and dispersion. Highly stylized diagrams are still used to explicate the concepts.

The curious fact is that though this structure is frequently postulated, it evidently has never been exploited. Typically, the identification of this structure is followed by identification of properties of random variables defined by cell counts for alternative formulations of stochastic spatial distributions. Once indentified, the studies fail to utilize the concepts of pattern and dispersion and do not indicate how cell counts are related to pattern and dispersion. Of course, it is not possible to use one measure to reflect these attributes of spatial distributions if, as presumed, they are independent.

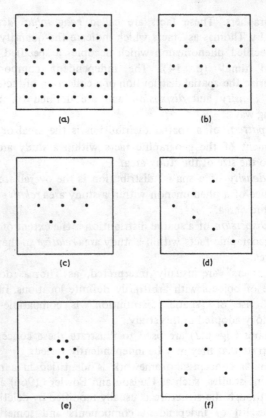

Figure 6.2 Thomas (1965) uses these diagrams to illustrate the following attributes of spatial distributions. Diagrams a and b have the same patterns and dispersions but differ in density. Diagrams c and d have the same densities and dispersions but differ in pattern. Diagrams e and f have the same densities and patterns but differ in dispersion.

There have been a few feeble attempts to isolate measures of these two components of spatial distributions. Dacey (1966a) uses the terms arrangement (instead of pattern) and dispersion. He shows, using stylized examples, that cell counts will not distinguish spatial distributions that intuitively are quite different. He suggests there is need to combine cell counts with contiguity measures, such as those súmmarized by Dacey (1964b), but does not indicate how these measures will reflect the two components of spatial distributions nor identify procedures to effect the combination. Another common suggestion is to utilize counts for several sizes of cells, but procedures for comparison of spatial distributions summarized by multiple cell counts are not clarified.

Several studies, such as Dacey (1966b), have suggested two measures of spacing that may be effective surrogates for arrangement and dispersion. It is convenient to identify these measures in terms of random variables defined for stochastic spatial distributions, but essentially similar spacing measures may be defined for map distributions.

Definition 5. Let $D = (R;\ \Omega_n;\ F)$ be the stochastic spatial distribution that is being analyzed. Let $D_1 = (R;\ \{\omega\};\ F_1)$ be a random spatial distribution defined by one object, distinguishable from those in Ω_n, uniformly located on R, and let $D_2 = D + D_1$ be the spatial distribution formed by the superpositioning of D and D_1. Two spacing measures are defined. For the spatial distribution D, let $Y_j, j \leqslant n - 1$, represent the distance from an object in Ω_n to the j-th nearest other object in Ω_n. For the spatial distribution D_2, let Z_j, $j \leqslant n$, represent the distance from ω to the j-th nearest object in Ω_n.

Because $\{Y_j\}$ is a measure of D that depends upon the relative locations of objects without regard to their locations with respect to the size or shape of R, $\{Y_j\}$ seems to serve as a surrogate for the arrangement of D. Because $\{Z_j\}$ is a measure of D that depends upon the locations of objects with respect to the size and shape of R, $\{Z_j\}$ seems to serve as a surrogate for the dispersion of D. The formulation of these components of spatial distribution is essentially operational: though the notions of arrangement and dispersion may be motivated by explanations that appeal to intuition, these components of spatial distributions are defined only in terms of the two spacing measures. These spacing measures are usually derivable for stochastic spatial distributions and are observable on map distributions.

While properties of these two spacing measures are now known for a variety of theoretical spatial distributions, these measures have not been integrated with descriptive and classificatory methods. The following examples, however, identify a property of these measures that, ·potentially, constitutes a fundamental criterion for comparison of spatial distributions.

Example 4. Consider the stochastic spatial distribution $D = (E; \Omega_\infty; F)$ and suppose F is specified in such a way that D is a poisson point process. Also, let $D_1 = (E; \Psi_\infty; F)$ be a poisson point process and $D_2 = D + D_1$. If, for D, Y_j is the distance from an object in Ω_∞ to the j-th nearest other object in Ω_∞, and if, for D_2, Z_j is the distance from a point in ψ_∞ to the j-th nearest object in Ω_∞, then Y_j and Z_j have the same probability distribution for each j.

Example. For the spatial distributions identified in Definition 5, suppose D is the random distribution identified in Example 1. The equivalence between Y_j and Z_j is not exact, but is a limiting property as $m(R)$ and n become large.

This relation between Y_j and Z_j evidently characterizes the random and poisson spatial distributions. However, it is not at all clear how the two measures may be integrated to yield information that is pertinent to the description and classification of spatial distributions. Moreover, experience with these two measures is insufficient to justify the assertion that arrangement (or pattern) and dispersion are independent and fundamental components of spatial distributions or that the two measures are adequate surrogates for these components.

Current status

The preceding examination of procedures available for the description and classification of spatial distributions justifies the assertion that there is not available a conceptual and methodological framework that structures the study of spatial distributions. Moreover, the available empirical and theoretical studies scarcely facilitate understanding of the properties and components that reflect similarities and differences in spatial distributions. Yet, a strong intuitive grasp of the notion of similarity in spatial distributions seems prerequisite to the development of descriptive, classificatory and analytic methods. Rather than discourse on the elusiveness of this seemingly simple notion, it is more effective to display problems in the context in which they occur and, thereby, explicitly identify some of the unresolved issues that confront attempts to formalize the concept of spatial distributions.

Examples

Diagrams are commonly used to facilitate the study of basic properties of spatial distributions. These diagrams are not particularly productive in that they utilize highly stylized, artificial examples, such as those illustrated in figs. 6.1 and 6.2, with no indication that there may exist underlying location rules. Instead of stylized diagrams, the examples used in this study are descriptions of stochastic spatial distributions. The use of these distributions not only admits a broader range of more interesting examples but also establishes that there are fundamental, though confusing, interactions between location rules and the spatial distributions they generate. The attempt has been to select the simplest spatial distributions that will carry the concepts that need to be clarified. So, while these examples serve the present purpose of illustrating basic issues, they are not necessarily indicative of the complexity of issues that actually confront the study of spatial distributions.

The three spatial distributions that serve as examples are the random, the Polya and the Polya-uniform spatial distributions. The random spatial distribution is specified by Example 1, while the Polya spatial distribution has been previously examined by Dacey (1969). For purposes of comparison, all three spatial distributions are defined in a consistent format. These examples are formulated in the context of a state divided into counties, though the state may be interpreted as any study region that is partitioned into non-overlapping subregions.

Example 5. The spatial distribution $D = (R; \Omega_n; F)$ is defined in terms of a collection $F(R, \Omega_n)$ of location rules that generate the locations in the state R of the n objects belonging to Ω_n. The state $R = (J, \{r_j\})$ has area $r = m(R)$ and is divided into J counties; the area of the j-th county is r_j and $r = r_1 + \ldots + r_j$. The (punctiform) objects are located one at a time in the state. The location rule F is specified in terms of the random variable $L_j(m)$, $1 \leqslant m \leqslant n$, $1 \leqslant j \leqslant J$. Let $L_j(m) = 1$ if the m-th object located in the state is located in the j-th county and otherwise $L_j(m) = 0$. Put $X_j(0) = 0$, $X_j(m) = L_j(1) + \ldots + L_j(m)$. In addition to objects being located one at a time, the location rule $F(R; \Omega_n)$ is defined in a way that satisfies the conditional probability

$$P\{L_j(m+1) = 1 \mid X_j(m) = k\} = \frac{r_j + kh}{r + mh} > 0, \quad \begin{cases} 0 \leqslant m \leqslant n-1, \\ 0 \leqslant k \leqslant m. \end{cases}$$

The spatial distribution D is summarized in terms of the random variables $\{X_j(m)\}$ and in terms of the following additional random variables. Let $X_A(m)$ represent the number of objects located in an arbitrary subregion $A \subset R$ at the time when the state contains m objects; by arbitrary is meant a subregion that is delimited without reference to the county structure of R. The spacing measures $Y_j(m)$ and $Z_j(m)$ represent object to j-th nearest object distance and sample point to j-th nearest object distance respectively (as in Definition 5) at the time the state contains m objects.

Definition 6. Suppose D satisfies the conditions of Example 5. If, in addition $h \neq 0$, then F is called the Polya location rule with parameter h and D is called the Polya spatial distribution with parameter h. The Polya location rule and spatial distribution are denoted by $F(p)$ and $D(p)$.

Example 6. Suppose the spatial distribution D satisfies the conditions of Example 5. In addition to these specified properties, suppose that the location rule F is defined in such a way that, given the county on which an object is located, then the object is uniformly located on the county.

Definition 7. Suppose D satisfies the conditions of Examples 5 and 6. If, in addition, h = 0, then F is called a random location rule and D is called a random spatial distribution. The random location rule and spatial distributions are denoted by $F(r)$ and $D(r)$.

Definition 8. Suppose D satisfies the conditions of Examples 5 and 6. If, in addition, $h \neq 0$, then F is called a Polya-uniform location rule and F is called a Polya-uniform spatial distribution. The Polya-uniform location rule and spatial distribution are denoted by $F(u)$ and $D(u)$.

Example 7. For the spatial distribution $D(p)$, the random variable $X_j(m)$ has the Polya distribution with

$$P\{X_j(m) = k\} = \binom{m}{k} \frac{B(k + u, m - k + v)}{B(u, v)}, \quad 0 \leqslant k \leqslant m \leqslant n,$$

where $u = r_j/h$ and $v = (r - r_j)/h$. It is not possible to derive the probability distributions of $X_A(m)$, $Y_j(m)$ or $Z_j(m)$.

Example 8. For the spatial distribution $D(r)$, the random variable $X_j(m)$ has the binomial distribution with parameters m and r_j. If

$m(A) = p$ and $m(R) = r$, then $X_A(m)$ has the binomial distribution with parameters m and p/r. In order to derive the probability distributions of $Y_j(m)$ and $Z_j(m)$, it is necessary to know the shape of R.

Example 9. For the spatial distribution $D(u)$, the random variable $X_j(m)$ has the polya distribution identified in Example 7. In order to derive the probability distributions of $X_A(m)$, $Y_j(m)$ and $Z_j(m)$, it is necessary to know the shape of R and the partition of R into counties, i.e., the shape of each county and its relative location in R.

The realizations $m(D)$ for the Polya and Polya-uniform spatial

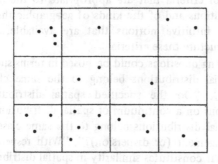

Figure 6.3 The two diagrams illus-
trate possible map representations
of the Polya and Polya-uniform
spatial distributions.

distributions differ in one important respect. For the Polya spatial distribution, locative information is restricted to the number of objects located in each county, while for the Polya-uniform (and random) spatial distributions, the locations of objects within counties are also known. Figure 6.3 illustrates possible realizations of the Polya and Polya-uniform spatial distributions.

Some questions about spatial distributions

Question 1. Does $D(p)$ correspond to intuitive notions of a "spatial distribution?"

Question 2. Are $D(p)$ and $D(u)$ "spatial distributions" of the same type? Is it necessary to differentiate them?

The following series of questions is directed to the comparison of properties of spatial distributions. It is a sterile and unproductive task to describe or classify spatial distributions whenever there is no clearly defined objective or purpose, for then there is no external evaluation of the utility or validity of a proposed description or classification. While the following questions involve description and classification without a specified objective, there is no intention to actually describe or classify spatial distributions. Instead, the examples and questions are intended to provoke consideration both of the kinds of criteria that are appropriate to the comparison of spatial distributions and of the kinds of geographic theories, spatial concepts and intuitive notions that are available, and may be invoked, to structure these criteria.

The following questions could be posed in terms such as "Do the specified spatial distributions belong to the same class of spatial distributions?", "Do the specified spatial distributions occupy nearby positions on a continuum of spatial distributions?", "Do the specified spatial distributions belong to the same class with respect to their arrangement (or dispersion)?", "With respect to intuitive notions of what constitutes similarity in spatial distributions, should a classification scheme assign the specified spatial distribution to the same class?", and "With respect to these intuitive notions, how should a classification scheme order or relate spatial distributions that are assigned to different classes?" To avoid lengthy lists of such questions, the study uses terms such as, "Are the specified spatial distributions similar or different?" However, "similar", "same" and "different" should not be given literal interpretations; instead, they are intended to direct attention to essential properties of spatial distributions and to the methodological structure that is involved with their description and classification.

Example 10. The spatial distribution $D_1(u) = (R; \Omega_n; F_1)$ is generated by the Polya-uniform location rule with parameter h_1 and the spatial distribution $D_2(u) = (R; \Omega_n; F_2)$ is generated by the Polya-uniform location rule with parameter $h_2 \neq h_1$. Random variables defined for D_1 and D_2 are indicated by a second subscript.

Question 3. Are the spatial distributions similar for all values of h_1 and h_2? Different? Does it matter that $V X_{j,1}(m)/E X_{j,1}(m)$ is less than unity for $h_1 < 0$ and greater than unity for $h_1 > 0$?

Question 4. What attributes, if any, discriminate the two spatial distributions? Do they reflect the location rule? Observable properties? Both?

Question 5. Are F_1 and F_2 defined by different location rules?

Example 11. Everything is the same as for Example 10 except that one n is changed to $m \neq n$.

Question 6. Are the spatial distributions similar? Different?

Question 7. What attributes, if any, discriminate the two spatial distributions? Is density a distinguishing attribute of spatial distributions? Is scale?

Example 12. The states R_1 and R_2 have different shapes but the same area. The spatial distributions $D_1(r) = (R_1; \Omega_n; F)$ and $D_2(r) = (R_2; \Omega_n; F)$ are random distributions.

Question 8. Do the spatial distributions have the same arrangement? The same dispersion? Are the spatial distributions similar? Identical?

Question 9. What attributes, if any, discriminate the two spatial distributions? Does similarity of spatial distributions depend upon shape of states?

Example 13. The states R_1 and R_2 are distinguishable in that they have different county structures and the counties have different areas. The spatial distributions $D_1(r) = (R_1; \Omega_n; F)$ and $D_2(r) = (R_2; \Omega_n; F)$ are generated by the Polya-uniform location rule with parameter h.

Question 10. Are the spatial distributions similar? Different?

Question 11. What attributes, if any, discriminate the two spatial distributions? What is the effect of differing shapes and sizes for the states?

Question 12. Is one location rule capable of generating different spatial distributions? Do distinguishable states always have different spatial distributions?

Question 13. Are the answers to the last three questions consistent with the answers to the three questions about Example 10? Is consistency a relevant issue?

Example 14. Suppose the states R_1 and R_2 are composed of the same J counties but differ in the arrangement of counties. The spatial distributions $D_1(u) = (R_1; \Omega_n; F)$ and $D_2(u) = (R_2; \Omega_n; F)$ are generated by the Polya-uniform location rule with parameter h.

Question 14. Are the spatial distributions similar? Different?

Question 15. What attributes, if any, discriminate the two spatial distributions? Do states with different internal structures always have different distributions?

Question 16. Will the random variables $X_{j,1}(n)$ and $X_{j,2}(n)$, $1 \le j \le J$, discriminate the two spatial distributions? Will $X_{A,1}(n)$ and $X_{A,2}(n)$?

Example 15. The states R_1 and R_2 differ in area but have the same relative county structure. The j-th county of R_1 has area r_j and the j-th county of R_2 has area br_j, $b > 0$. The spatial distribution $D_1(u) = (R_1; \Omega_n; F)$ is generated by the Polya-uniform location rule with parameter h, and the spatial distributions $D_2(u) = (R_2; \Omega_n; F)$ is generated by the Polya-uniform location rule with parameter bh.

Question 17. Are the spatial distributions similar? Different?

Question 18. What attributes, if any, discriminate the two spatial distributions? Is scale a distinguishing attribute? Is scale an attribute of $D_1(u)$?

Question 19. Are $D_1(u)$ and $D_2(u)$ defined by different location rules?

Question 20. Suppose the subregions A_1 and A_2 have the same area. Given that A_1 and A_2 have corresponding locations in R_1 and R_2, it can be shown that $X_{A,1}(n)$ and $X_{A,2}(n)$ have the same probability distributions. Does this fact alter the answers to the three previous questions?

Question 21. Do the answers to the preceding four questions imply that different location rules generate similar spatial distributions? Can they?

It is frequently asserted that there exist spatial distributions $D_1 = (R; \Omega_n; F_1)$ and $D_2 = (R; \Omega_n; F_2)$ such that F_1 and F_2 are "obviously" different but any pairs of comparable measures for D_1 and D_2, such as counts of objects located in corresponding subregions, have identical probability distributions. The justification for this assertion is usually an adaptation of one of the following two properties.

Property. Many common random variables may be generated both as the sum of a random number of random variables and as mixtures of random variables.

Example. The Neyman type A probability distribution is the sum of a poisson distributed number of random variables having a common poisson distribution and also is a mixture of poisson distributions by the poisson distribution.

Property. Many common random variables may be generated by different models.

Example. The negative binomial distribution is a mixture of poisson distributions by the gamma distribution and also is the number of trials to the k-th success in a sequence of Bernoulli trials.

These properties indicate that different models may generate the same random variable. Though some of these random variables are frequently used to summarize cell counts on observed map distributions, it does not follow that there exist spatial distributions D_1 and D_2 such that F_1 and F_2 are different location rules but D_1 and D_2 are summarized by identically distributed random variables. The next two examples illustrate the types of models that are typically used to substantiate the assertion that different location rules generate indistinguishable spatial distributions.

Example 16. The spatial distribution $D_1(u) = (R; \Omega_n; F_1)$ is defined by the Polya-uniform location rule with parameter h. The number $X_{j,1}(n)$ of objects located in the j-th county has the Polya distribution, identified in Example 7, with parameters n, $u = r_j/h$ and $v = (r - r_j)/h$. The spatial distribution D_2 is defined by letting $F_2(R; \Omega_n)$ be the (random) location rule such that each object is located in county j with independent probability Q_j. The value of Q_j is not fixed, but is the realization of a random variable that has the beta distribution with parameters $a_j > 0$ and $b_j > 0$, or

$$P\{Q_j \leqslant t\} = \int_0^t x^{a_j-1} (1-x)^{b_j-1} \frac{dx}{B(a_j, b_j)}.$$

Then, $D_2 = (R; \Omega_n; F_2)$ is the spatial distribution generated by locations on R of objects in Ω_n when the probability that an object locates in a county is the realization of a random variable. Given that $Q_j = x$, the random variable $X_{j,2}(n)$ has the binomial distribution with parameters n and x. So,

$$P\{X_{j,2}(n) = k\} = \int_0^1 P\{X_{j,2}(n) = k \mid Q_j = x\} x^{a_j-1}(1-x)^{b_j-1} \frac{dx}{B(a_j, b_j)}$$

$$= \int_0^1 \binom{n}{k} x^k(1-x)^{n-k} x^{a_j-1}(1-x)^{b_j-1} \frac{dx}{B(a_j, b_j)}$$

$$= \binom{n}{k} \frac{B(k+a_j, n-k+b_j)}{B(a_j, b_j)},$$

or $X_{j,2}(n)$ has the Polya distribution with parameters n, a_j and b_j. When $a_j = u$ and $b_j = v$, then $X_{j,1}(n)$ and $X_{j,2}(n)$ have identical probability distributions. For the following questions, suppose these equalities are satisfied.

Question 22. Are the spatial distributions similar? Different?

Question 23. Are D_1 and D_2 different location rules?

Question 24. Are two different location rules capable of generating similar spatial distributions? Identical spatial distributions?

Quesiton 25. Does D_2 satisfy intuitive notions of a properly defined location rule?

Question 26. Is it possible to construct an experiment that generates a map representation m of D_1? Of D_2? If it is not possible to construct an experiment that generates $m(D_2)$, is D_2 a spatial distribution?

Question 27. Need every spatial distribution have a map representation?

Question 28. Is it possible to interpret the parameters a_j and b_j in the context of location theory? Should it be?

Question 29. Is the interpretation of parameters simplified if it is specified that all J counties of R have the same area?

Example 17. The counties of R have equal area and the number J of counties is large. For $D_1 = (R; \Omega_n; F_1)$, F_1 is the Polya-uniform location rule with parameter h, and D_1 is summarized by the counts of objects located in a random sample of counties. Since all counties are identical, the J random variables $X_{j,1}(n)$ are identically distributed as, say, the random variable $X_1(n)$, which has the Polya distribution with parameters n, $u = s/h$ and $v = (r - s)/h$. For $D_2 = (R; \Omega_n; F_2)$, F_2 is the (random) location rule such that each object is located in county j with independent probability p_j. The value of p_j varies from county to county, and let this variation be approximated by the beta distribution with parameters u and v. Suppose D_2 is summarized by the counts of objects located in a random sample of counties, and let $X_2(n)$ represent the number of objects in a randomly selected county. Then,

$$P\{X_2(n) = k\} = J^{-1} \sum_{j=1}^{J} P\{X_{j,2}(n) = k\}$$

$$= J^{-1} \sum_{j=1}^{J} \binom{n}{k} p_j^k (1 - p_j)^{n-k}$$

$$\simeq \int_0^1 \binom{n}{k} x^k (1 - x)^{n-k} x^{u-1} (1 - x)^{v-1} \frac{dx}{B(u, v)}$$

$$= \binom{n}{k} \frac{B(k + u, n - k + v)}{B(u, v)},$$

which is the Polya distribution with parameters n, u and v.

Question 30. Are the spatial distributions similar? Different?

Question 31. Are F_1 and F_2 defined by different location rules?

Question 33. Does F_2 satisfy intuitive notions of a properly defined location rule? What conditions should a location rule satisfy?

Question 33. In what ways does this location rule F_2 differ from the rule F_2 of Example 16? Is this rule interpretable in the context of location theory?

Question 34. Is it possible to construct an experiment that generates a map representation m of D_2?

The following questions refer to the diagrams in fig. 6.2 that Thomas uses to explicate the notions of pattern and dispersion.

Question 35. With respect to Thomas' definitions of pattern and dispersion, do the descriptions of these diagrams, given in the caption, correspond to intuitive notions of similar patterns? Similar dispersions?

Question 36. Do the operational definitions for arrangement and dispersion provide a basis for the classification of these spatial distributions? If so, do they yield classifications that correspond to intuitive notions of similar spatial distributions? Should the results of a classification have intuitive appeal?

Question 37. Are the similarities and differences of these spatial distributions comprehended within the concepts of arrangement and dispersion? These concepts plus density? Are these answerable questions?

A methodological context is provided for the following, more general, questions by the accompanying quotations from the recent geographical literature. These quotations do not, of course, use a consistent terminology.

... Analysis of the nature of two-dimensional distributions in the abstract should be able to furnish a theoretical framework with capacity to illuminate actually observed distributional patterns and space relations. (Ackerman 1958, 28).

A typical theoretical situation in geography is described usually by way of patterns. Patterns are morphological laws. (Schaefer, cited by Bunge. 1966, 207)

There is still very little evidence in the behavioral studies that geographic process theory can be formulated in such a way as to yield statements of spatial structure as logical "output". (King 1969, 594)

Question 38. Is the geographic study of stochastic spatial distributions focused upon location rules or upon measures of properties that may be observed on map distributions? Or both? Or somewhere else?

. . . When patterns and relationships are investigated in terms of basic categories, unifying principles are revealed. (Warntz 1958, 453)

Dacey's models (of spatial distributions) are verifiable, but they tell us little or nothing about process. (Webber 1971, 27)

. . . The numerical is less indispensable than the geometrical; for all which is distributed has geometrical form. And what are the laws about the change of form, structure and scale of the distribution other than statements about the causal meaning of a geometric characteristic of the distributed phenomena? (Tschierske 1961, 108)

These abstract models must then be confronted with the reality of an actual landscape. (Christaller 1964, 257)

Question 39. Is the geographic study of map distributions focused upon the identification of location rules that are capable of generating the observed distribution or upon properties that may be used for classification? Or both? Or somewhere else?

Question 40. Does similarity of map distributions refer to similarity of location rules and of observed properties? Only to similarity of observed properties? Or to something else?

It is the constraints of the situation in combination with the specific makeup of the behavioral matrix . . . that largely determines the degree to which the corresponding distribution . . . will appear to be more clustered, dispersed or evenly spaced than random. (Pred 1967, 31)

. . . Geography . . . must pay attention to the spatial arrangement of the phenomena in an area and not so much to the phenomena themselves. Spatial relations are the ones that matter. (Schaefer 1953, 228)

Question 41. Is the geographic study of a map distribution $M(R; W_n)$ independent of the attributes that determine membership in W_n? Independent of the size of R? Independent of the shape of R?

It is frequently asserted that the important and critical aspect of research is problem formulation rather than problem solving, in that it is more difficult to identify the right question than to obtain the correct answers to given questions.

Question 42. Has this study posed questions that contribute to clearer comprehension of the concept of spatial distribution?

References

ACKERMAN, E.A. (1958) Geography as a fundamental research discipline; *University of Chicago, Department of Geography Research Paper No. 53.*

BUNGE, W. (1966) Theoretical Geography; *Lund Studies in Geography, Series C. No. 1 (2nd ed.).*

CHRISTALLER, W. (1963) no title; *Ekistics* 16.

DACEY, M.F. (1964a) Modified poisson probability law for point pattern more regular than random; *Annals of the Association of American Geographers* 54, 559-65.

DACEY, M.F. (1964b) A review of measures of contiguity for two and k-color maps; *Northwestern University Department of Geography Research Paper No. 8.* Reprinted in BERRY, B.J.L. and MARBLE, D.F. (Eds.), *Spatial Analysis,* (Prentice-Hall, Englewood Cliffs, N.J.), 1968, p.479-95.

DACEY, M.F. (1966a) A county seat model for the areal pattern of an urban system; *Geographical Review* 56, 527-42.

DACEY, M.F. (1966b) A probability model for central place locations; *Annals of the Association of American Geographers* 56, 549-68.

DACEY, M.F. (1969) A hypergeometric family of discrete probability distributions: Properties and applications to location models; *Geografiska Annaler* 1, 283-317.

HARTSHORNE, R. (1939) The Nature of Geography; *Annals of the Association of American Geographers* 29, Nos. 3 & 4 (special issue).

HARVEY, D. (1969) *Explanation in Geography;* (Edward Arnold, London).

HUDSON, J.C. and FOWLER, R.M. (1966) The concept of pattern in geography; *University of Iowa, Department of Geography Discussion Paper No. 1.*

KING, L.J. (1969) The analysis of spatial form and its relation to geographic theory; *Annals of the Association of American Geographers* 59, 573-95.

PRED, A. (1967) Behaviour and Location; *Lund Studies in Geography, Series B, No. 27.*

ROGERS, A. (1969) Quadrat analysis of urban dispersion: 1. Theoretical techniques; *Environment and Planning* 1, 47-80.

SCHAEFER, F.K. (1953) Exceptionalism in geography: A methodological examination; *Annals of the Association of American Geographers* 43, 226-49.

THOMAS, E. (1965) A structure of geography: A proto-unit for secondary schools; *High School Geography Project, Association of American Geographers,* (University of Colorado, Boulder, Colorado).

TSCHIERSKE, H. (1961) Raumfunktionelle prinzipien in einer allgemeine theoretischen Geographie; *Erkunde* 15, 92-110.

WARNTZ, W. (1958) Geography at mid-twentieth century; *World Politics* 11, 442-54.

WEBBER, M.J. (1971) Empirical verifiability of classical central place theory; *Geografiska Annaler* 3, 15-28.

PART III

Environmental

7 · Geography as human ecology

RICHARD J. CHORLEY

Although December 1972 marks the fiftieth anniversary of Barrows' (1923) important address to the Association of American Geographers, it may seem strange that one should adopt his title for an essay written half a century later. However, the notion that geography can be justified as an application of the ecological model to man in society is more deeply ingrained now than ever before. Man's growing concern with such matters as conservation, pollution, quality of life and the like seem to have revived a vision of his returning to some folk-remembrance of harmony and nature, and the application of sophisticated techniques to the analysis of the ecosystem (Patten 1971) has encouraged the fusion of this vision with fashionable quantitative scholarship. The purpose of this essay is to examine one aspect of the ecological approach to geographical methodology, namely to what extent it provides a unifying link between current human and physical geography.

Rousseau's paradox

It is more than two centuries since Rousseau in his *Discours sur les sciences et les arts* (1750) formulated a view of human development in terms of the paradox that mankind deteriorates as material civilization advances. Nevertheless this paradox increasingly perplexes those geographical methodologists who view the discipline in terms of man-land relationships, and it is no accident that Glacken's (1967) important treatise on the relationships of man to the habitable earth terminates in the mid-nineteenth century. Glacken saw this hiatus largely in philosophical terms wherein the long-dominant concept of a 'designed earth', which forms the thesis of his work, was to be successively modified by the theories of evolution and ecology. However, the classical studies in regional geography demonstrate that it was only before the middle of the nineteenth century, at least in the western world, that man's relationship to the habitable earth could be even approximately interpreted within an ecological model. Although the term ecology and the geographical appropriation of it are early twentieth century in date, the Industrial Revolution and its attendant economic and social ramifications had long before made the ecological model out of date as a geographical template for large areas of the world, as Vidal de la Blache clearly showed in his work on Eastern France (1917, 120-1 and 126-7; Wrigley 1965, 9-11). As an historical theory it may be appropriate to associate the concept of ecology with the mid-nineteenth century idea of evolution, but as a retrospective geographical model its utility had then already been overtaken by events in the more advanced industrial counties. It is thus important to recognise that there must always be a clear distinction between the historical formulation of scientific models and their appropriate historical application. Changes in geographical outlook are also found to result from tangible social and economic developments, and those who wish to apply the concept of geography as human ecology face serious problems in attempting to treat contemporary social man in the same manner as a plant or animal community. In Ackerman's (1963, 431) words, the concept is increasingly too ambitious.

As Durkheim (1906, 132) pointed out, Rousseau demonstrated that, if all that comes to man from society were peeled off, there would remain nothing but a creature reduced to sense experience and more or less undifferentiated from the animal. Such a creature might fit well into an ecological man-land model, but one cannot peel off the increasingly complex socio-economic controls over the spatial manifestations of man's relations with the earth. We can no longer model for ourselves 'geographical man' as an extension of Rousseau's 'natural man'. The model of the natural physical and biological environment as a wise and benign teacher using its negative-feedback mechanisms to admonish the avaricious excesses of a human society which is becoming more and more 'unnatural' as its numbers and complexities proliferate may have been a useful maxim for Edwardian boy scouts but it is not of much assistance in modern social man's search for his developing terrestrial role. Geographers have long made a habit of attacking new problems with outmoded models, and the belief that a man-embracing ecological approach can wholly reconcile the dichotomy between human and physical geography is an attempt to confront our current environmental problems with the visions of Wordsworth and Emerson. Our responsibilities to the physical environment are much more positive than this, for the latter is fast becoming more and more man-dominated — or even man-created.

The ecological model may fail as a supposed key to the general understanding of the relations between modern society and nature, and therefore as a basis for contemporary geographical studies, because it casts social man in too subordinate and ineffectual a role. Although much interesting and vital research is at present being conducted into the ecosystem aspects of some more isolated societies, into less advanced agricultural communities and with regard to the origins of agriculture, it is difficult to see how this analogy can profitably be extended to form a model which will encompass the activities of an important and rapidly-expanding proportion of social and economic life styles. The idea that flows of capital investment, population, technological information, generated energy, water, and the like, together with such constraints as involve interest policies and the mechanisms of group decision-making, can be reduced to comparable units so as to be structured into energy linkages similar to those of ecosystems is clearly an illusion, despite implications to the contrary (e.g. Linton 1965). Human geography is no simple extension of biogeography, things have gone too far for that — if they were ever so. Even the Garden of Eden had its entrepreneur. Man's relation to nature is increasingly one of dominance and control, however lovers of nature may deplore it. If the proponents of geography as a scholarly discipline wish to continue to reflect the relationships between society and nature they cannot afford to adopt models which ignore the glaring probability that this relationship is one which exists between an increasingly-numerous, increasingly-powerful and progressive, if capricious, master and a large, increasingly-vulnerable and spitefully-conservative serf. Of course it would be a foolish master who did not diligently study the characteristics of his subordinate in order to so moderate his own actions as to extract the maximum efficiency from his employee and to keep him fit for future work. But man has other employees, some less tangible but nevertheless, real, such as considered. Man's relation to nature has emerged not as one of self-interrogation, nor even of simple dialogue, but of a conference with others present and in which social man increasingly has the right of veto.

For me, geography concerns itself with the tangible spatial manifestations of the continuing intercourse between Man and his habitable environment. It is thus something less than 'the science of location', as is demonstrated by the fact that nearest-neighbour analysis was first applied to the location of forest trees and the Neyman Type A distribution was developed to describe the distribution of larvae hatched from eggs. It is also something less than the 'study of the earth as the home of Man', for 'home' implies more than the location of rooms and the operation of a suitably-scaled spatial socio-economic domestic system. Philosophers, poets and writers have also had much of importance to say regarding Man's relation with his earthly abode, but it is just a futile to pretend that the regional novel or the nature poem is a contribution to geography as it is to claim geographical domination over all locational studies at every spatial scale. Geography, however, will remain an inherently spatial discipline whose ultimate concern is with those landscape features produced and modified by the corporate actions of social man, together with those features of the 'natural' environment which occur within spatial scales, or which change within temporal scales, appropriate in exerting some influence or constraint over the spatial manifestations of man's activities. The problem which geographers face, now as always, is to decide how deeply it is necessary to involve themselves with spatially-intangible processes in order to sustain a meaningful explanation of these spatial features of landscape (Langton 1972).

Resource management

It may seem ironical that, after almost two decades of the so-called 'quantitative revolution', the discipline should still find its methodological development bound up with one of the basic philosophical problems which preoccupied the medieval world, namely, to what extent is it proper to regard Man as a part of Nature or as standing apart from it? Of course, the relationship between Man and the 'natural world' has long provided a recurring theme in geography, but so far as practical attitudes and patterns of work are concerned geographers have tended to regard the matter as having been settled more than a hundred years ago. Glacken (1967, 549-50) has shown how, by the nineteenth century, the older teleology of the 'web of God's design' had begun to decline, and has suggested how the work of Darwin was to replace it by the considerations of environmental adaptations and interrelationships (the 'web of Life') which culminated in the twentieth-century concept of the ecosystem. Many geographers have implied that man in society might be viewed as forming part of a complex ecological system, most effectively Stoddart (1967). For example, although George Perkins Marsh (1864) was impressed by man's ability to dominate the landscape, his was still essentially an ecosystem view of man's unity with his environment. His version of the ecological role of man finds its modern counterpart in the ecosystem management approach (e.g. Van Dyne 1969). Although detailed analyses of the operations of simple ecosystems have been achieved only recently, the idea of the ecosystem has long had an impact on geographical theory, not least on the scope and concept of physical geography (Chorley 1971, 92). As has been shown elsewhere (Chorley 1971; Chorley and Kennedy 1971), the developments of the geographical revolution have placed such strains on the traditional bonds between physical and human geography as to make necessary radically new conceptual approaches to the subject, all of which have had the effect of blurring the existing intra-disciplinary distinctions.

The concept of geography as human ecology has thus had a long history of implicit application to our discipline and has been broadly viewed as providing one of the reasonable integrating frameworks for a subject which traditionally deals with the spatial interrelationships between the socio-economic and physico-biological aspects of reality. The limitations of such an application of this model to the organization of the subject matter of contemporary geography are, however, becoming rather apparent. As the discipline has been forced by the realities of the modern world to become more and more concerned with socio-economic matters, the 'natural' environment has become a more and more subsidiary part of the total system of study. Indeed, the latter is increasingly recognised to have been replaced by a largely man-made environment which is subordinated to the socio-economic environment to a much greater extent than other organic life forms are able to control their environments. A bigger difficulty lies in the essential structure of the more primitive ecosystems studied by botanists and biologists, in that they are dominated by negative feedback loops and homeostatic considerations to do with energy flows, food chains, and the like. Systems involving social man have, by contrast, stronger inbuilt positive feedbacks, implying a time-dependence which often becomes the dominant feature of the system operation. Despite modern work showing that the distinctions between negative and positive feedback are less sharp than was hitherto believed (e.g. Milsum 1968), the problem still remains an important methodological one for geographers. Nothing illustrates the limitations of the ecosystem approach to much of the geography of the long-populated regions of the world than the differences between the ecosystem 'management' approach on the one hand and more advanced socio-economic 'planning' on the other. Management, in this sense, is seen to involve the manipulation of the equilibrating operations of the ecosystem so as to achieve higher levels of production, generally within the existing structure of the system. Ecosystem management is therefore primarily associated with the maximization of existing productivity, the minimization of wastage by the adoption of a suitable harvesting strategy, pest control, or the scientific cropping of native flora and fauna (Watt 1968; Van Dyne 1969). On the other hand, much advanced socio-economic planning concerns itself with the replacement of one ecosystem by another or, at the very least, with the impelling of the ecosystem through of trajectory of non-recoverable states. It is significant that when ecological managers refer to the creation of new ecosystems it is commonly within a context, for example, of decrying attempts by early settlers to cultivate semi-arid or humid tropical areas in ways similar to those of temperate cultivators. Clearly, the ecosystem model is of geographical significance in so far as Man can be considered to operate in the same manner as other life forms, the model is inappropriate in so far as Man stands apart from Nature. Thus the application of the ecosystem concept to the organization of geographical studies raises two fundamental issues: Are human systems so much more complex than the biological parts of ecosystems as to severely restrict the geographical application of this model? And, if so, what is the role of Man as a controller of the system with which he interacts?

There are many difficulties facing geographers who have interests in both the natural environment and in many spatial socio-economic processes. One of the currently most intractable is that, having been imbued with the ecosystem model with its emphasis on balance, equilibrium, cycling and stability, geographers are increasingly faced with the methodological necessity of also accommodating active control involving the impelling of systems on time trajectories through sequences of states each different, probably non-recoverable and presumably ever-more adapted to the evolving needs of man in society. The matter is made more complex by the dynamic nature of these longer-term social needs in terms of population totals and their evolving social goals, and by the increasing difficulty in predicting what will be required of the earth as a home suitable for man's occupancy. In short, geographers are being faced with the basic problem of modelling systems which are stable in the short term under negative-feedback mechanisms, yet are capable of long-term changes under the positive-feedback mechanisms involved in economic and social tendencies. If one were to put one's finger on one of the really basic theoretical obstacles to geographical progress at present it would be that geographers have been so indoctrinated with such homeostatic concerns as characterize the ecosystem approach, with all that this implies, that they are disoriented by the stark possibility that irresistable tendencies already exist which are causing social man to manipulate his resources, including space, in ways which cannot be rationalized solely in terms of short-term equilibrium.

Another departure from any simplistic model of social man's relation to his habitable environment is furnished by the recognition that optimization models aimed at resource and space allocation, however apparently complex and all-embracing, have not in practice commonly formed the bases for significant planning decisions. This has caused some geographers to concern themselves increasingly with the mechanics of group decision-making regarding the exploitation of the environment and this, in turn, has resulted in a growing interest in the political implications of decision-making institutions.

'Political geography' has thus acquired a novel and relevant focus which is very much bound up with basic political beliefs which question *by* whom environmental decisions involving limited resource allocation (including space) are being made, and for the benefit *of* whom.

If political geography has recently acquired a new lease of life, no branch of the discipline has more exemplified the continuing difficulties associated with the geographer's traditional concern with the spatial artifacts of man's activity than does historical geography. In its categorizing of morphological assemblages of landscape and their mutual interpretation in time slices, some of the methods of classical historical geography were closely allied with those of denudation chronology. The mutual relationships between morphological artifacts of the operation of past spatial systems can only be clearly established by linking them by means of the cascades of energy, finance or information which sustained their functional association. It is therefore no accident that the most attractive studies of such historical geography are those in which the artifacts of man's spatial activity are rationalized in terms of adequate records of such cascades (Baker 1972). Such data are clearly most available for the more recent past, and when they exist for the more distant past, commonly in the form of financial returns, they breathe life into such morphological studies.

It is, however, less easy to dismiss the morphological residua of past systems than modern proponents of selective, systematic, regional geography have assumed. Just because it is not possible to rationalize and structure these morphological features within currently-operating spatial systems, it does not mean either that their existence can be ignored or that they have no significant, if indirect, effects on the operation of contemporary systems of interest to geographers. Simply on account of their occupancy of often key locations, historical artifacts may continue to play an important contemporary role.

Control systems

The application of systems analysis to geography (Stoddart 1967; Chorley and Kennedy 1971; Langton 1972), greeted by some as a conceptual breakthrough and by others as a jargon-ridden statement of the obvious, has at least served to highlight and rationalize some of the important and long-continued methodological difficulties which geographers have faced. It has also allowed scholars to identify within such theoretical models as that of the ecosystem those essential basic and inter-disciplinary features which, though the dictates of the model, pervade patterns of thought and research goals.

Traditionally, geography was not well prepared to adopt a viable systems approach in that geographers had tended to concentrate upon the artifacts of the operation of spatial systems, rather than on the system themselves (Langton 1972). The *landschaft* approach, the concern with the fossil forms produced by past erosion (denudation chronology), and with the landscape remains of past systems of human occupancy have traditionally tended to throw the emphasis away from the study of the operation of spatial denudational, social and economic systems. However, human geography was able to exploit the systems approach more immediately and satisfactorily than other branches of the discipline, although not specifically in those terms, because its contemporary concern enabled its artifacts to be interpreted most directly in terms of the operating systems which produced, sustained and transformed them. It is thus no accident that locational analysis and urban geography were the first inheritors of the systems legacy. Although other branches of geography, as a whole, have been less systems-susceptable than these, it is very significant that the parts which recently have developed most rapidly (i.e. hydraulic geometry and nineteenth-century historical geography) are those too, which exhibit most obviously the relationships between operating systems and their artifacts.

The idea that the subject matter of geography somehow involves the tangible morphological features of the 'natural' and the man-made landscape is still as important today as it was half a century ago under its *landschaft* designation (Sauer 1925). The conceptual clarity of systems analysis has given further proof of the difficulties long-associated with attempts to treat these morphological features, or artifacts, of spatial process-response systems to some extent independently of the cascades of mass, energy and information which produced them, sustain them, or transform them at the highly-variable rates appropriate to their system relaxation times. Exclusive concentration on morphology, whether directed to the natural or man-made landscape, can only result at best in description or in crude attempts at 'explanation' effected by means of correlations between morphological variables. It is now clear that any success in the latter endeavours has been due solely to the extent to which a given independent morphological variable is expressive of storages or flows forming parts of the cascades. In the past, of the major branches making up the established discipline of geography, geomorphology has been the least conceptually hindered by this problem because its concentration on water, an observable and measureable part of the landscape, as the assumed major cascade of terrain process-response systems. Recently, however, the importance of solar energy in terrain cascades is coming to be recognised as a limitation of earlier work. The cascades which support those morphological features of the man-made environment which form the objects of study in human geography are far less tangible than those upon which classical geomorphology has relied. Flows of energy and information, storages of skills and capital, and regulators governed by complex mass attitudes and operated by means of complex decision-making bodies all represent intangibles (in a *landschaft* sense) on which the morphology of landscape is built. It is in this sense that some geographers have seen a useful analogy between human geography and plant and animal ecology in that the latter, too, represent partly-inanimate process-response systems supported by variable photo- and bio-chemical cascades of a sophisticated character.

The development of the systems approach to geography has naturally resulted in a critical analysis of its methodology. We are told that systems analysis has not been shown to be particularly useful in analysing change in human geography (Langton 1972); that systems are too complex and structurally-rich to be conveniently analysed, particularly when the inputs are continually changing; and that real systems are neither strictly equilibrium nor timebound, but that they lurch from one state of disequilibrium to another. In short, we are led to believe that the real-world systems cannot readily be modelled in a form which neither reproduces their interlinked equilibrium operations and changes through time, nor permits the adequate testing of such models. In the light of these comments it is interesting that at least one author has pointed out that, in the study of ecosystems, evolution itself can be understood in the framework of systems analysis (Margalef 1968). At this level of generality the analogy ceases to be helpful to methodologists in human geography, however, for plant and animal communities may be too simple and too much a prey to negative feedback processes to permit the ecosystem to become a useful general model for geographical systems.

'Control' in the ecosystem sense thus merely implies the manipulation of the negative-feedback loops in order to stabilize the system operation at some optimum state (Milsum 1966). Although much of control engineering literature is concerned with similar regulator systems or servomechanisms, engineering systems analysis is turning increasingly to the study of systems in which the component controlled variables are kept in close correspondence with a frequently-changed reference variable ('Follow-up systems') (Harrison and Bollinger 1969), or when the operation of the system itself generates information which, in turn, causes the aims of the system operation to be changed ('Learning control systems') (Greensite 1970). In all fairness, however, one must add that the theoretical development of time-directed engineering control systems is at present much less operational in terms of its possible applications to geographical problems than is the timeless equilibrium ecosystem model or the engineering servomechanism. However, it is clear that the past emphasis on the latter, rather than on both types of system, has introduced many difficulties in terms of the analysis of geographical systems involving socio-economic as well as physical process-response components (Chorley and Kennedy 1971).

Nowadays virtually every suggestion having to do with the restructuring of geography and with its need to increase its relevance with respect to current environmental problems involves, explicitly or implicitly, the application of some measure of systems analysis. It is interesting that up to the present the use of systems analysis in structuring and solving problems of interest to geographers has been largely restricted to multiple-regression type models (Chorley and Kennedy 1971), although somewhat more dynamic situations have been modelled in terms of hydrological systems and ecosystems. Despite this, it is clear that many geographers are becoming impatient with the systems approach because, although it seems extremely attractive from a methodological point of view, the spatial interactive situations of interest to geographers appear on the other hand to be so complex, so rich in connectivity, so dominated by unknown and largely non-linear transforms, and so complicated by lags and variable thresholds as to defy as yet an all-embracing systems attack. Again, the fact that it is so much more easy to construct equilibrium models than those exhibiting a time-trajectory has limited the appeal of the systems approach to the increasing number of geographers who are concerned with prediction and planning. This bias has inevitably thrown the methodological emphasis in systems applications in geography upon the equilibrium models of ecology and of classic economics, rather than those embodying strong positive-feedback loops as in some advanced engineering control systems and in the multiplier-accelerator economic model. When attempts have been made to construct dynamic systems of some complexity (e.g. by Forester 1969), the sweeping prior assumptions regarding variable relationships and the mechanisms of changes through times have naturally laid them open to apparently damning criticism. Nevertheless the fact remains that, despite its present limitations, both the frame of mind induced and the research prospects held out by systems analysis require that this approach must be seriously explored as the major methodological effort of geography during the next few years.

The kind of geographical methodology which, in my view, is increasingly necessary is analogous to that used in analysing a man-machine system. In the geographical context the 'machine' is made up of those systems structures of the physical and biological environment which man is increasingly able to manipulate (either advertently or inadvertently) together with the physical artifacts of man's activity—buildings, lakes, polluted air, and the like. Although it has been convenient in the past to depict such physical and biological systems as compartmented cascades (Heinmets 1969; Chorley and Kennedy 1971, 77-125), it is clear that one of the most significant conceptual extensions of environmental systems analysis lies in the complete adaptation of such models to stochastic processes (Matis and Hartley 1971). In passing, it is as well to note that the distinction between 'man' and 'natural environment', so central to the thinking of lovers of nature and conservationists, is not a clear one and is daily becoming more confused. Substantial portions of the world have landscapes which are not 'natural' in the sense that man has already impelled them through non-recoverable thresholds. Even on a global scale man is increasingly influencing the solar energy budget and the cycling of water, sediment and chemicals. 'Man' in the context of geographical man-machine analogy is composed of a complex of interlocking socio-economic systems which operate on the above 'machines' by means of decision-making mechanisms of great complexity and bias, not without their own important stochastic elements. What can be conceived of as the environmental 'control' exercised in this context thus represents the manifestation of a systems-complex of great sophistication with which the geographer must inevitably concern himself.

It is by his very increase in numbers, resulting in the growth of material strength, proliferation of competitive demands and in the increasing complexity of organization which so impressed Rousseau, that social man is being more and more set apart from the physical and biological environment. Competition to exploit, control and consume all manner of terrestrial resources, including space, is impelling man to continually extend his environmental dominance. Many prognosticators have expressed the very real fear that the continued unbridled operation of positive socio-economic feedback loops, involving the exponential growth of population and of industrial capital, must lead to their ultimate arrest by the lagged operation of the negative feedback loops of pollution, resource exhaustion and famine (Forrester 1971 Meadows et. al. 1972, 156-7). However, it is apparent that, far from encouraging man to relax his overall environmental grasp, such tendencies may result in proposals for increased planning and control. It seems clear, therefore, that geographers might temper their preoccupation with the ecological type of model in favour of the application of that of the control system, so that, adapting Glacken's terminology, 'God's design', having been replaced by 'Nature's design', may be in turn supplanted by 'Man's design'. Needless to say, it is much too early to pass judgement on the ultimate wisdom and quality of the latter. It is clear, however, that social man is, for better or worse, seizing control of his terrestrial environment and any geographical methodology which does not acknowledge this fact is doomed to inbuilt obsolescence.

References

ACKERMAN, E.A. (1963) Where is a research frontier?; *Annals of the Association of American Geographers*, 53, 429-440.

BAKER, A.R.H. (1972) Historical geography in Britain; Ch. 5 in BAKER, A.R.H. (Ed.), *Progress in Historical Geography*; (David and Charles, Newton Abbot), 90-110.

BARROWS, H.H. (1923) Geography as human ecology; *Annals of the Association of American Geographers*, 13, 1-14.

CHORLEY, R.J. (1971) The role and relations of physical geography; *Progress in Geography* 3, 89-109.

CHORLEY, R.J. and KENNEDY, B.A. (1971) *Physical Geography: A Systems Approach*; (Prentice-Hall, London).

DURKHEIM, E. (1906) Détermination du fait moral; *Société Française de Philosophie, Bulletin* 6, 113-138.

FORRESTER, J.W' (1969) *Urban Dynamics*; (M.I.T. Press, Cambridge, Mass.), 285p.

FORRESTER, J.W. (1971) *World Dynamics*; (Wright-Allen Press, Cambridge, Massachusetts).

GLACKEN, C.J. (1967) *Traces on the Rhodian Shore*; (University of California Press, Berkeley and Los Angeles), 763p.

GREENSITE, A.L. (1970) *Elements of Modern Control Theory*; (Spartan Books, New York), 886p.

HARRISON, H.L. and BOLLINGER, J.G. (1969) *Introduction to Automatic Controls*; 2nd Edn. (International Textbook Co., Scranton, Pa.), 460p.

HEINMETS, F. (Ed.) (1969) *Concepts and Models of Biomathematics: Simulation techniques and Methods*; (Marcel Dekker Inc., New York).

LANGTON, J. (1972) Potentialities and problems of adopting a systems approach to the study of change in human geography; *Progress in Geography* 4, 125-179.

LINTON, D.L. (1965) The geography of energy; *Geography* 50, 197-228.

MARGALEF, R. (1968) *Perspectives in Ecological Theory*; (Chicago), 111p.

MARSH, G.P. (1864) *Man and Nature: or Physical Geography as Modified by Human Action*; Lowenthal, D. (Ed.), Harvard University Press, 1965).

MATIS, M.H. and HARTLEY, H.O. (1971) Stochastic compartmental analysis: Model and least squares estimation from time series data; *Biometrics* 27, 77-102.

MEADOWS, D.H., MEADOWS, D.L., RANDERS, J. and BEHRENS, W.W. (1972) *The Limits of Growth*; (Earth Island Ltd., London), 205p.

MILSUM, J.H. (1966) *Biological Control Systems Analysis*; (McGraw-Hill, New York), 466p.

MILSUM, J.H. (Ed.) (1968) *Positive Feedback*; (Pergamon Press, Oxford), 169p.

PATTEN, B.C., (Ed.) (1971) *Systems Analysis and Simulation in Ecology*; Academic Press, New York and London).

ROUSSEAU, J-J. (1750) Discours sur les sciences et les arts; In MASTERS, R.D. (Ed.), *The First and Second Discourses*, (St. Martin's Press, New York).

SAUER, C.O. (1925) The morphology of landscape; *University of California Publications in Geography* 2, No. 2, 19-54.

STODDART, D.R. (1967) Organism and ecosystem as geographical models; In CHORLEY, R.J. and HAGGETT, P. (Eds.), *Models in Geography*, (Methuen, London), 511-548. (An excellent analysis.)

THOMAS, W.L. (Ed.) (1956) *Man's Role in Changing the Face of the Earth*; (Chicago), 1193p.

VAN DYNE, G.M. (1969) *The Ecosystem Concept in Natural Resource Management*; (Academic Press, New York and London), 383p.

VIDAL de la BLACHE, P. (1917) *La France de L'Est*; (Armand Colin, Paris), 280p.

WATT, K.E.F. (1968) *Ecology and Resource Management*; (McGraw-Hill, New York), 450p.

WRIGLEY, E.A. (1965) Changes in the philosophy of geography; Chapter 1 in CHORLEY, R.J. and HAGGETT, P. (Eds.), *Frontiers in Geographical Teaching*, (Methuen, London), 3-20.

8 · Energy-based climatology and its frontier with ecology

F. KENNETH HARE

... the heat and water balance of the earth's surface is,
... as a rule, the main mechanism that determines the
intensity and character of all the other forms of exchange
of energy and matter between ... the climatic, hydrologic,
soil-forming, biologic and other phenomena occurring on
the earth's surface ... the study of the natural heat and
water balance of the earth's surface, the determination of
its role in natural geographical phenomena and the working
out of methods for purposeful change are the scientific
foundation for the development of the theory and practice
of the main types of natural heat and water ameliorations,
i.e., the problems of the transformation of nature."

M.I. Budyko, I.P. Gerasimov 1961

The above statement of a widely-held view of contemporary physical geography is written in language that contains the presuppositions of physics, geography and Marxism. It is confident in tone because the authors had behind them the authority of the Presidium of the Academy of Sciences of the U.S.S.R., which in 1954 defined as one of the most important problems of Soviet science the study of "the heat and water régime of the earth's surface, its role in the dynamics of natural processes and methods of transformation for practical purposes." Both versions stress the practical usefulness of such study, a viewpoint by no means confined to the Marxist countries.

Synthesis is easy to announce, but hard to pull off. I have been waiting for the underlying unity of physical geography to assert itself ever since in the 1930's (when Grigor'yev was attempting such a synthesis in the Soviet Union) I heard Wooldridge and Tansley put forward essentially similar views. Tansley's ecosystems were far less deterministic in conception than the Budyko-Gerasimov statement, but he also saw the natural surface of the earth as made up of physical-biotic unities, to be comprehended as systems. Wooldridge taught Davisian geomorphology as a series of orderly responses to climatic control, and taught climatology as the study of atmosphere and land surface linkages. I remember his prediction that within the lifetime of his hearers (though not his own) the small sub-disciplines that studied the earth's surface would merge. So far it has not happened. If anything the number of sub-disciplines has increased, and so has the range of esoteric jargons. Nevertheless, such a convergence must be sought, if we are to avoid the Tower of Babel. Climatology, at least, has moved strongly in the direction that Budyko and Gerasimov foresaw, and helped propel it.

I am hopeful that the crystallization of climatology around energy considerations will help lead towards the desired synthesis. In the following pages I have tried to show how the crystallization has led to a better understanding of the relation of climate to soil, plant and animal life, and have concluded that this is a more promising avenue of exploration at present than the links with geomorphology.

Trends in climatology

The shift towards the understanding of the surface energy and mass exchanges, especially inland over vegetated surfaces, has been the main achievement of the past twenty years of climatology, and I have no doubt that it will continue. The preoccupations of earlier years — for example, the attempt to classify climates and establish climatic regions, in the manner of Köppen, or the statistical analysis of standard meteorological data, in Conrad's fashion — have yielded to a thorough-going effort to understand what happens when water, carbon dioxide, heat and momentum are exchanged or transformed at the air-land and air-sea interface. This is the territory of micrometeorology, or microclimatology in Geiger's sense. But macroclimatology has followed suit, in that we now try to see the large-scale climatic distributions in terms of the same parameters. My own present work consists of attempting to understand the water and heat balances of North America in valid physical terms, the criterion of validity being that they can be substantiated experimentally on the small scale, or deduced from physical theory.

The history of this shift has been effectively summarised by Miller (1965), and there is no point in repeating what he has written.

Soviet scientists generally attribute the movement to Voyeykov's work early in this century, and it was certainly the work of Budyko, Director of the Voyeykov Main Geophysical Observatory in Leningrad, that stimulated the conversion of English-language climatologists to this approach. His great *Heat Balance of the Earth's Surface* (1956, translated by N. Stepanova in 1958) and the *Atlas of Heat Balance* (1955) were unique documents at the date of their appearance, and still have few imitators.

That Budyko's initiative was so quickly effective was the result of the preparing of the ground by a handful of western geographers, most notably Leighly and Thornthwaite, both students of Sauer. Thornthwaite's second system of classification (1948) stood very much on a threshold. It was still aimed at the moribund objective of classification, and its central parameter, potential evapotranspiration, was still wholly empirical. But behind these traditional qualities stood Thornthwaite's insistence on rationality, and on the quantitative assessment of the roles of solar energy, transpiration, dry matter production, soil moisture storage and moisture surplus. The paper will be remembered, not for its working method, but for the shrewd insights it contained.

Geographers should recall, however, that many other specialists have contributed to this field, most of whom would be bewildered to hear their work called climatology (at least in the western world). It is useful to classify these other contributions according to the scientific method involved:

(i) *the micrometeorological method*, tracing back to the interest of fluid dynamicists, meteorologists and others in problems of atmospheric turbulence, diffusion and stability. The work of G.I. Taylor, Prandtl, Sutton, Kolmogorov and others established a pattern of experimental techniques that provided, and still provides, the physical insights needed into the nature of the earth's surface and planetary boundary layers. This branch of experimental physics is very much concerned with "on-site" effects, and is completely detached from the climatological viewpoint, to which it is nevertheless essential. In post-war years the volume of such work has been large; of special importance in the present context has been that of Monin and Obukhov in the

Soviet Union, and the brilliant group in C.S.I.R.O., in Australia, with Priestley as the prime mover (see Priestley 1959, for a summary). Closely related in method and attitudes, but more specifically addressed towards surface exchanges, one can identify

(ii) *the microclimatological method*, practised by a number of post-war physicists, soil scientists, agricultural meteorologists and a growing number of physical geographers. In this case the approach is to apply the techniques of micrometeorological measurement and boundary layer theory, together with the related parts of soil physics and plant physiology, to the exchange and transformation processes for mass and energy over specific natural surfaces. In Britain Penman and Monteith (both very much physicists) have raised this method to high levels. Elsewhere, Priestley, Swinbank, Dyer, Taylor, McIlroy and Slatyer in Australia, and Tanner, Lemon, van Bavel, Knoerr, Mather, King and Pelton in North America are names that spring to mind, though the list is lengthening rapidly. Among geographers, the group at McMaster University (Davies, Rouse and their colleagues) are applying similar methods.

From these groups we have gradually acquired an understanding of how energy is transformed at the land surface, of how water moves upwards and downwards through the plant canopy and the soil, and of how carbon dioxide is assimilated during photosynthesis and released again during respiration (though this involves as well the quite different physiological skills of men like Gaastra and Lieth). A third group, who are heavily dependent on the foregoing, practise

(iii) *the hydrological method*, where the primary skills lie in the capacity to measure streamflow, ground-water storage and movement, and other related aspects of the hydrologic cycle. Very much the domain of obscure engineering methods until recent decades, scientific hydrology and hydrometeorology have today the aspect of physical sciences in the same tradition as climatology itself. Here again Russian initiative, in the hands of such workers as Glushkov, Drozdov and L'vovich, has seen this as a move towards a general understanding of physical geography. By a happy chance, similar initiatives in the United States by

Leopold, Langbein and others, were undertaken in an effort to relate streamflow to landforms, in such a way as to bring geomorphology into closer relation with climatic processes.

Climatology on the planetary scale is now often seen as the attempt to explain world climatic distributions in the light of what these methods of on-site research have taught us. In this attempt we are still badly hampered by the commitment of the world's meteorological services to those observations applicable to the dynamical techniques of forecasting. Since radiation has not been thought one of these, the most fundamental climatological measurements of all, those of the solar and terrestrial radiative fluxes, are badly neglected outside a few countries, notably the Soviet Union and Canada (which is the only country to publish the observations of her growing radiation network on an hourly basis). This neglect has been especially damaging in the United States, where, however, the new era of satellite observation is beginning to alter attitudes.

There are many other examples of our unbalanced monitoring of vital environmental parameters. We have continuing CO_2 mixing ratio determinations by infra-red gas analyser techniques only from a handful of stations: the oldest, the Mauna Loa series, dates from 1957. There is no international turbidity network, although some plans are been effected, in spite of growing evidence of catastrophic increases (Peterson and Bryson 1968; McCormick and Ludwig 1967). Large-scale climatology has had to be largely based on ingenious transformations of data acquired for other purposes.

Annual balance relationships

The first, and still unique, world synthesis by Budyko (1956, trs. 1958) is essentially an attempt to overcome these difficulties. Soviet priorities had in any case been different, and Budyko inherited an existing tradition of radiation, streamflow, soil temperature and moisture measurement. The book, and the Atlas that preceded it, are essentially based on ingenious extrapolation and approximation techniques. These are not beyond criticism, as Monin's (1964) attack and Budyko's (1964) reply attest. Most of the world distributions drawn by Budyko are still the only estimates available.

They include charts of the chief components of the annual surface energy and water balance, expressed in the equations (neglecting storage change)

$$R = H + LE; \qquad P = N + E \quad \dots \dots \dots \dots \dots (1)$$

where   H = convective heat (enthalpy) flux
            E = evapotranspiration, N = run-off, P = precipitation
            L = latent heat of evaporation
and   R, the net radiation or radiation balance, is defined by

$$R = I(1-a) + R\!\downarrow - R\!\uparrow \quad \dots \dots \dots \dots \dots (2)$$

where   I = incident global solar radiation on a horizontal surface,
            a = albedo, and
            $R\!\downarrow$ and $R\!\uparrow$ are the downward and upward long-wave radiative
            fluxes at the surface

World charts of annual values of $I, I(1-a)$, H and LE are derived from direct observations of P and I alone. Comparable charts of N have been prepared by L'vovich (1962). I G Y data are incorporated into revised versions by Budyko, Yefimova, Aubenok and Strokhina (1962) for R, H and LE, and a revised edition of the Atlas (1963) does likewise.

These charts are to be regarded as brilliant improvisations in the face of an accute dearth of observations. Monin (1964) took special exception to the derivation of H, and some of the others involve approximations or assumptions that do not satisfy micrometeorologists. In effect, however, Budyko's estimates represent the triumph of climatological faith over micrometeorological pessimism. If one needs estimates of the basic parameters of the world's heat and moisture exchanges, one must approximate where one cannot measure. The only imperative is that one approximate as closely as one can the real physics underlying the obscurity.

In his treatment of the physico-geographical zonation of the earth, Budyko makes use of a set of dimensionless ratios

$$C = \frac{N^*}{P^*}; \quad F = \frac{E^*}{P^*}; \text{ and } D' = \frac{R^*}{LP^*} \quad \dots \dots \dots \dots (3)$$

where the asterisks indicate annual mean values, R^* is specifically for a well moistened surface, and C, F, and D' are respectively the run-off ratio, the evaporation ratio and the radiational index of dryness. Clearly if storage changes are negligible $F = 1 - C$. Budyko then puts

$$1 - C = F = f(D') \qquad \qquad \qquad \qquad (4)$$

determined empirically as a regression relationship and not as a monovalent equation (Monin 1964).

In a treatment certain to become a classic in its field, Lettau (1969) introduces the Bowen ratio of the annual means ($B = H^*/LE^*$), and defines the dryness ratio, D, identically with Budyko's radiational index of dryness, except that R is the observed annual mean net radiation over the actual surface. He writes

$$(1 + B)(1 - C) = D \qquad \qquad \qquad \qquad (5)$$

This remarkably simple statement of relationship between the basic parameters of the moisture and energy balances is in fact the general balance equation of surface climatology. It is devoid of numerical coefficients and arbitrary constants, since it merely expresses the conservation principles for moisture and energy. With its help equation (4) of Budyko becomes

$$C = 1 - \frac{D}{1 + B}; \text{or } F = \frac{D}{1 + B} \qquad \qquad \qquad (6)$$

which is deterministic, and hence monovalent, unlike (4). It states that the Bowen ratio, the evaporation ratio F (the evapotranspiration as a fraction of the annual precipitation) and the dryness ratio, D, form a conjugate set. If any two are known, so is the third. If water supply is unrestricted, and B tends to zero, F approaches D, and hence D'. D is then the ratio of potential water demand (R^*/L) to supply P^*.

Sellers (1965) has called attention to the formal relationship of Thornthwaite's moisture index m, to D' (and hence to D). For a moist surface, with $B = 0$, it follows that

$$m = 100(D'^{-1} - 1) \qquad \qquad \qquad \qquad (7)$$

Much the same parameter, D' or D, hence crops up in each of the best-known attempts to regionalise the earth on the basis of the physical climate of the surface.

In equation (6) B is not a known function of D, and solutions for C and F must hence depend on further observational input, or on assumed forms for B. Budyko (1958) gives as a solution

$$C = 1 - [D' \tanh D'^{-1} (1 - \cosh D' + \sinh D')] \quad \cdots \quad .(8)$$

which Lettau (1969) fortunately simplifies to

$$C = 1 - \tanh D \quad \cdots \cdots \cdots \cdots \cdots .(9)$$

One can thus derive a set of parameters — the run-off, dryness and Bowen ratios — that describe to a reasonable approximation the annual régime of energy and water inputs, run-off and evapotranspiration. There remains a scatter due to seasonal rhythms, which express themselves by varying Bowen ratios. All the parameters are rational derivations from simple conservation principles, and of the relationships required for practical application only (8) and (9) are empirical regressions. It has been shown that the Thornthwaite, Budyko and Lettau formulations of the effective dryness of a climate are nearly equivalent. Clearly these ideas have great strength. They carry authority because they are not in the least obscure or empirical except for the C to D relation, and this will eventually be overcome by a parameterization of the Bowen ratio (Lettau, *personal communication*).

In order to apply these methods trustworthy observations are needed of at least two of the raw variables, precipitation, radiation, run-off, and evapotranspiration. In practice only precipitation and run-off are generally available, and then with often dubious reliability. Solar radiation may also be available, from which the net radiation can be approximated. If two raw variables alone are in hand, equations like (8) or (9) are needed to achieve a solution. Budyko's work in the Soviet Union used precipitation, radiation and run-off data, and from this he published charts of the radiational index of dryness for the whole country. He extended the analysis world-wide by using equation (8).

Hare (1971) has published charts of the dryness, run-off and Bowen ratios of the annual mean for Canada, and has noted a poor fit in northern areas. Hare and Hay (1971), using monthly net radiation determinations by Hay (1970) for Canada and Alaska, have showed that the annual fields of evapotranspiration and convective heat flux derived from observations of precipitation and run-off appear spurious. The authors ascribe much of the lack of fit to faulty snowfall measurement.

A virtue of these methods is that they are highly explicit, and they are unlikely to hide errors in observations. Their very simplicity and directness of derivation makes them potentially as useful in detecting such errors as in defining the gross heat and moisture balance climates of the earth.

Biosphere relations

The idea just formulated — that the effective climate of the earth's surface depends on the supply of water and energy, and the way in which they combine in the hydrologic cycle — is far from new. What is new is our growing ability to dispense with past approximations, and with the essentially arbitrary indices so typical of the older climatology. As we have moved in this direction — towards energy-centred methods — so also have the ecologists. This is the age of trophic-dynamic ecology, to use Lindeman's term. If synthesis is feasible, we might look for it here.

Lindeman's initiatives, taken in the late nineteen-thirties (he died in 1942 at the age of 26), were based on the energy-conservation principle. Drawing mainly from the previous conceptual frameworks of Thienemann (1918, 1926, 1939), Elton (1927) and Tansley (1935), as well as on the Clementsian school, Lindeman treated natural ecosystems on the basis of the capacity of their primary producers (photosynthetic plants) to capture part of the incident solar and atmospheric energy, and to wrap it up in the dry vegetable matter that would then yield it to the grazing and decay food webs. Within each ecosystem successive trophic levels were established, each typically involving energy about one order of magnitude lower than the level above it. Respiration from plants, animals and soil completed the cycle by returning heat and carbon dioxide to the atmosphere. Succession was seen as the movement towards a stable climax biomass in which photosynthesis and respiration had come into equilibrium. Lindeman's final posthumous synthesis (1942) proved to be one of the most influential scientific papers of this century.

There is a large volume of published literature in this field, and a substantial part of the work of the International Biological Programme aims at establishing quantitatively the present biomass within the major ecosystems, their structure, function (in energetic terms, as well as in others) and dynamics. It is now possible to state in broad terms the range of net dry matter production typical of the major ecosystems. The photosynthetic relation is

$$6\,CO_2 + 12\,H_2O \xrightarrow{\text{689 kcal}} C_6H_{12}O_6 + 6\,O_2 + 6\,H_2O \quad . \quad . \quad (10)$$

for which it follows that the gross production of a kilogram of glucose requires nearly 4,000 kcal. The synthesis of tissues requires the immediate metabolic consumption of part (or the order of a quarter to a half) of the glucose (and hence release of energy). The net production of dry matter is the gross production by relation 10 less this metabolic loss.

The efficiency and level of net production are clearly related to the energy climate. It is well-known that only the visible spectrum is effective in photosynthesis, though the temperature of the tissues involved depends on the net total energy exchange, radiative and convective, with the environment. Specific climatic influences have been discussed in detail by Gaastra (1959, 1963), Lieth (1963) and Monteith (1966). The latter was able to define a model giving gross potential photosynthesis in sugar beet as a function of global solar radiation (all wavelengths) and day length. Adopting Monteith's values, and assuming further (i) that growth ceases below 6C, and (ii) that respiration losses are governed by a simple linear relation with air temperature, Jen-Hu Chang (1970) has published world charts of potential net photosynthesis. Recent results (Paltridge 1970) suggest that maximum net production in a wide range of species and climates is close to 320 kg ha^{-1} day^{-1}, well above reported world averages.

If the conversions of energy reported by these workers are weighted for areal extent and for the preponderance of infertile sites (including the near-sterility of the sea), the net world annual photosynthetic energy conversion is seen to be much less than one per cent of the net radiation. The earth's food-webs hence exist at energy densities at least two and sometimes three or four orders of magnitude below those characteristic of the ambient climate. Only a tiny fraction of the incident solar energy is actually trapped and stored. At first sight it is hence easy to conclude that the biosphere's functions are trivial on the energy scale of climatology.

This is indeed true at sea, where production is very low, and where there is no transpiration. But over a humid land surface at climatological Bowen ratios below 0.60 (the land-surface average), most of the net radiation goes into transpiration, which is necessary to the photosynthesis. We hence arrive at the familiar conclusion that the strongest atmosphere-biosphere links *over land* are physiological, and are concerned with the use of energy at the plant surface to evaporate water. From this we can reason that the anatomy of the individual plant, and the structure and physiognomy of the vegetation, are certain to be related to the energy balance climatology. For it is these aspects of ecosystems that relate most closely to the transpiration process.

I cannot yet point, however, to a single complete quantification of these links (but see Paltridge 1970, for a gallant attempt). It is clear that there are broad regularities in the relation between ecosystem structure and ambient climate. We have been trying for over a century to classify these regularities, by showing that certain climatic parameters have distributions like those of certain vegetation boundaries. The near-coincidence of the 10 C July mean daily air temperature isotherm and the arctic tree-line is an example, and Köppen's system of classification assumed that there are several similar spatial correspondences. In 1973 such ideas seem flimsy and unsatisfying.

Present-day ecological research into productivity lays greater emphasis on biomass and dry matter production than on structural linkages with the atmosphere. There are, however, many attempts to quantify the links in energy-based terms. The Russian physical geographers have again been very zealous. In Table 8.1 I present, slightly modified, Grigor'yev's scheme of geographical zonality.

It is apparent that Grigor'yev's correlations are in fact another example of fitting specific forms of natural vegetation (physiognomically identified) to the energy and moisture regimes. The

Table 8.1. Grigor'yev's system of geographical zonality

Radiational index of dryness (D')	Net radiation range 0 – 50 (kcal/cm²/year)	Net radiation range 50 – 75 (kcal/cm²/year)	Net radiation range Over 75 (kcal/cm²/year)
<0.2	Arctic desert	—	—
0.2-0.4	Tundra & forest tundra	Sub-tropical hemi-hylaea with swamps	Equatorial forest swamps
0.4-0.6	Northern & middle taiga	Sub-tropical rain forest	Very swampy equatorial forest
0.6-0.8	Southern taiga & mixed forests	same	Moderately swampy equatorial forest
0.8-1 (optimal)	Deciduous forest & forest-steppe	same	Equatorial forest, passing into lighter tropical forest and wooded savannahs.
1-2	Steppe	Hard-leaved sub-tropical forests and scrub	Dry savannah
2-3	Temperate semi-desert	Sub-tropical semi-desert	Desert savannah (tropical semi-desert)
Over 3	Temperate deserts	Sub-tropical deserts	Tropical desert

Source: Grigor'yev (1961)

climatic parameters used, however, are R and D', as defined on page 176 above. These are fully rational and deterministic in character. The substantial success of Grigor'yev's scheme confirms that the gross physiognomy of the world's ecosystems is rationally related to the energy climate.

Similar relationships have been established by Gerasimov (1961) between genetic soil types and the energy and moisture regimes. Table 8.2 shows his world scheme. It is not easy to relate his terms to detailed western soil classifications, but the general force of his work is clear. Like Grigor'yev, Gerasimov correlates soil type with the radiational index of dryness, D', but his energy correlation is with the net radiation of the warmer season. For the latter he uses as an index accumulated surface temperatures (T_s) above 10 C. Budyko (1958, fig. 55) shows that to a good approximation

$$R^+ \text{ (for season above } 10°C) = \Sigma(T_s - 10)/100 \qquad \ldots (11)$$

though this correlation tends to break down in low latitudes.

Recent Russian research has considerably extended these attempts at synthesis. Bazilevich and Rodin (1971) have published world maps of total phytomass, values ranging from below 2.5 tons ha^{-1} in the high arctic tundra to over 500 tons ha^{-1} in tropical rainforest; their world estimates of annual net production vary from below 1.0 ton ha^{-1} to near 50 tons ha^{-1} over the same range. Drozdov (1971) has correlated net production with both R and D', showing a significant correlation in each case, but tending to refute Grigor'yev's choice of D' = 1 as optimal for growth; highest growth came at lower values. For good moistening conditions (D' = 0.4 to 1.4) his curves indicate the following normal relations between R and net production:-

Annual net radiation (kcal cm^{-2})	Annual net production (tons ha^{-1})	Energy conversion* (%)
15	2	0.6
30	8	1.2
45	14	1.4
60	21	1.6
75	29	1.8

*assuming an energy content of dry tissues of 4.5 kcal g^{-1}

Table 8.2 Gerasimov's scheme for the climatic correlations of the world's main genetic soil types

$(\Sigma T_s-10)/100 = R^+$	Radiational Index of Dryness			
	0.5-1.0	1.0-1.7	1.7-2.5	Over 2.5
10-20	Podzols of northern & middle taiga	–	–	–
20-28	Turf-podzols of southern taiga	–	–	–
28-50	Brown forest soils of broadleaf forests	Chernozems of steppes	Chestnut soils of dry steppes	–
50-80	Red earth & yellow earths of moist subtropical forests	Cinnamon-brown soils of dry subtropical forests / Reddish black soils of subtropical forests	–	Gray-cinnamon brown soils and gray-earths of subtropical steppes and semi-deserts
80-100	Laterites of equatorial forests	Red soils of tall grass tropical savannahs	Red-cinnamon brown soils of dry subtropical forests	Red-brown soils of steppe-like and desert-like tropical savannahs

Source: Gerasimov (1961)
N.B. Where R^+ is <10, tundra soils occur.

The relation is nearly linear, production increasing equatorward at roughly 1 ton ha^{-1} per 2 kcal cm^{-2} added radiation.

The Grigor'yev and Gerasimov syntheses, valuable though they are because of their explicit dependence on the surface energy and water balances, still fall short of their own objectives, well summed up by Grigor'yev himself as

. . . . . a comparative study of the structure, dynamics and development of the geographic zones. . . . .; an investigation . . . . . of the interrelationships and interdependences, and of the interchange of matter and energy between the components of the geographic envelope; a determination on that basis of the magnitudes and qualitative differences among the annual productivities of the bio-mass of the plant cover and animal life . . . . . as indices of the level of bio-energy produced by the biotic components of the geographic envelope . . . . . and the determination, on the basis of all these data, of the basic laws of the structure, dynamics and development of the geographic environment. (Grigor'yev 1961, 5-6)

"Laws", at least to the western scientist, mean a good deal more than simple classifications and enumerations of what happens in specific categories of energy-moisture combination. In complex fields like physical geography and synecology "laws" yield place to intellectual comprehension of how ecosystems function. This involves knowing how the various mechanical, radiative, material and informational links actually work. Systems ecologists are working in this direction now, but I have not yet seen models that satisfactorily incorporate the physical climate.

Elsewhere (Hare 1950, 1954) I tried to show that the structure and physiognomy of the Boreal Forest formation were rationally related to the Thornthwaite potential evapotranspiration function (PE), which can be looked on as an attempt to predict the quantity R*/L from air temperature data. Recently, with J.C. Ritchie, I have returned to this problem (Hare and Ritchie 1972), armed with Hay's (1970) comprehensive analysis of the various radiative fields over Canada. What is now apparent is that the links between northern vegetation (and animal life) and climate are two-way links. The surface climate depends on the surface characteristics, which in turn depend on the ability of vegetation to grow under marginal climatic conditions. Specifically the demonstrable linkages include the following:

(i) the arctic tree-line, the northern limit of the boreal woodland, and the northern closed-crown forest line appear to follow specific isolines of the annual net radiation, or, still more closely, specific isolines of growing season net radiation, and the length of the season with mean daily temperature above 0C (Hare 1970; Hare and Hay 1971). The biotic limits concerned are strictly physiognomic, and appear to reproduce themselves all round the pole under widely different floristic situations;

(ii) there is a northward gradient of standing phytomass from high values along the northern forest boundary to vanishingly low values along the arctic coast-lines of Canada and Siberia. (Bliss 1962; Andreyev 1966; Alexandrova 1969). This gradient, like the physiognomic divisions, appears to be related to the net radiation field, as would in any case be expected;

(iii) a key role in the survival of the biota during winter is played by the snow-cover. This cover is thin, heavily-drifted and dense (>0.33 relative density) on the tundra, deep, less dense and more uniform in the boreal woodland and the forest-tundra. (Formozov 1946; Pruitt 1970; Hare 1971). The snow-cover provides shelter for plants and animals, and heavily influences soil temperatures and permafrost distribution.

Hence it would be easy to conclude that the energy and snowfall regimes are the controls, and the vegetation and animal life the responses, within the northern ecosystems. This is fallacious, because

(iv) the distribution and physical properties of snow-cover are governed largely by the structure of the vegetation. In the forest-tundra, for example, it is common for the discontinuous shrub layer (usually dwarf birch) to act as the stabiliser for snow, which would otherwise drift. Snow depth and density are functions, not only of the rate of snowfall and wind velocity, but of the effective roughness of the surface — which is largely governed by the structure of the vegetation;

(v) the net radiation regime is heavily influenced by the albedo (i.e., reflectivity) of the surface for solar radiation. Again this is a function of the composition and structure of the vegetation.

Snow-covered tundra has an albedo of 0.6 or more even in May and June, the months of maximum solar irradiation of the surface. The nearby boreal closed-crown forests have a low albedo (below 0.3, or even below 0.2 when very dense) even with deep snow-cover, because the effective surface is the dark-green canopy, which hides the snow. The closed-crown forest zone is even visible, because of this effect, on present-generation meteorological satellite photographs. The properties of the vegetation hence strongly influence the northern distribution of net radiation, and of snow-cover.

These are system relations of the classic kind, and parallel relations must exist in all ecosystems. The "laws" of physical geography, to revert to Grigor'yev's phrase, have to account for complex interrelationships of this sort. The future study of biosphere-atmosphere links, so vital to the geographer, will be dominated by efforts to demonstrate how such systems function. The establishment of spatial correlations, like those of Grigor'yev and Gerasimov, are valuable first steps in this direction, but hardly take us beyond the starting line.

Nevertheless, one can only feel optimistic for the future. Climatological and ecological methods are indeed converging, and we are moving in the direction of closer understanding. The climatologist's interest in vegetation differs, of course, from the present preoccupations of most ecologists. He is interested in the structure of the vegetation, because it is so vital to the turbulent exchange processes in the surface boundary layer, and to the radiative exchanges at the surface. He is led to believe that the uniformity of structure over great areas is a reflection of a highly specific set of interdependencies with climate. Ecological attention to such questions is not yet close. But considerable progress is likely in the next few years, as a common language emerges.

Conclusion

Space and my own limitations prevent me from exploring the geographical frontiers of climatology much further. It is my impression that the most rapid advances have been made, and will continue to be made, in the direction of the biosphere. Geographers have everything to gain from (and much to offer to) these advances. The present concern about environmental questions makes them highly topical.

Geomorphology seems to have offered a less easy path for integration. In spite of the spread of quantitative, physically-inspired methods, and a growing preoccupation with process, geomorphologists have not put their discipline on an energy-balance basis to nearly the same extent, and probably with good reason. A few words in defence of this conclusion are all that I can attempt.

There is first the obvious fact that fluvial processes tend to be dominated by extreme events, rather than balance relationships. It is well-documented observationally that the overwhelming preponderance of erosive and transporting work is done in the brief periods of peak flow. Though the mechanics of this work are thoroughly Newtonian, the way it gets done puts geomorphologists more closely in touch with the hydrologist and hydraulic engineer than with the climatologist. Stochastic methods and extreme-value theory are closer to the reality of geomorphic process than is energy-balance climatology.

Secondly, the geomorphic time-scale is enormously long, and the speed of processes incredibly slow, by comparison with most climatic processes. In principle this is not necessarily damaging; as the European school of climatic geomorphology made clear long ago, the shifting equilibrium of world climate must be reflected somehow in the land-forms of the present-day world. Nevertheless the kind of climatology discussed in this chapter concerns itself in practice with time-scales so brief that negligible geomorphic work is done while they go through their paces. It is the weathering process that most clearly depends on the energy and water balances, but even this proceeds at a snail's pace by comparison, for example, with the pace of succession in most ecosystems.

Glaciology is in a very different situation, and glacial meteorology is nothing but a special case of energy and moisture balance climatology. The same is true of the energy relations of the arctic pack-ice. Whether one calls these activities climatology is of no account. What matters is that they form common intellectual ground with what is called climatology in this paper.

What of the future? I am quite sure that it will see the rise to supremacy of two related, highly quantitative fields. The first is what we can loosely call systems methods. Ecosystems modelling is already an exciting reality, and it will in time extend itself to include the physical climate in a genuinely two-way fashion (which is not achieved now). The second, and in my judgement the key, development will be the completion of what my friend Heinz Lettau calls *climatonomy*, a word we could well adopt.

What Lettau has set out to do, and will achieve, is to establish a genuine, versatile theory of climate, capable of exact mathematical expression, and dispensing altogether with the empiricism, the arbitrary constants and the fudge-factors of the earlier decades of this century. He has already attempted to model evapotranspiration (Lettau 1969) and the radiative energy balance (Lettau and Lettau 1969), and has declared his intention of extending this climatonomy to a parameterization of biosphere exchanges, and of atmospheric chemistry. In my judgement Lettau's initiatives are likely to be as revitalizing for climatology as Lindeman's was for ecology. Much of this paper is written in his language.

Moreover, Lettau's ideas are leading us rapidly in the direction espoused by Budyko and Gerasimov in 1961. It is in the direction, not of Babel, but of a common, quantitative, theoretically-based language. That, in my judgement, is what physical geography needs. They lead us, also, in the direction that all science aims at, towards prediction. The future habitability of the planet may well depend on our success.

References

ALEXANDROVA, VERA D. (1969) The vegetation of the tundra zones in the U.S.S.R. and data about its productivity; in *Proceedings of the Conference on Productivity and Conservation in Northern Circumpolar Lands*, International Union for the Conservation of Nature, Publication No. 16 (new series), paper No. 9, 93-114.

ANDREYEV, V.N. (1966) Peculiarities of zonal distribution of the aerial and underground phytomass on the East European Far North; *Bot. Zh.* 51, 1401-11 (cited by Alexandrova, 1969; not examined by author).

BAZILEVICH, N.I. and RODIN, L. Ye. (1971) Geographical regularities in productivity and the circulation of chemical elements in the earth's main vegetation types; *Soviet Geography*

XXII(1), 24-53. (Original in *Obshchiye teoreticheskiye problemy biologicheskoy produktivnosti*, Leningrad: Nauka, 1969, 192 pp.)

BLISS, L.C. (1962) Adaptations of arctic and alpine plants to environmental conditions; *Arctic* 15, 117-44.

BUDYKO, M.I. (1955) *Atlas of Heat Balance*; (Chief Administration of the Hydrometeorological service, Leningrad), 41 pp. (Revised edition, 1963).

BUDYKO, M.I. (1958) *The Heat Balance of the Earth's Surface*; (translated by N. Stepanova from original dated 1956) (U.S. Weather Bureau, Washington), 259 pp.

BUDYKO, M.I. (1964) Reply (to attack of A.S. Monin, see below); *Soviet Geography* V, 18-31. (original in *Izvestiya Akademii Nauk SSR* seriya geografichiskaya 1964 (no. 1), 101-12).

BUDYKO, M.I. and GERASIMOV, I.P., (1961) The heat and water balance of the earth's surface, the general theory of physical geography and the problem of the transformation of nature; *Soviet Geography* II(2), 3-11, (Papers of the Water-Heat Balance Symposium, Third Congress of the Geographical Society of the U.S.S.R.).

BUDYKO, M.I., YEFIMOVA, N.A., AUBENOK, L.I. and STROKHINA, L.A. (1962) The heat balance of the surface of the earth; *Soviet Geography* III, 3-16. (Original in *Izvestiya Akademii Nauk SSSR*, seriya geograficheskaya 1962 (1), 6-16).

CHANG, Jen-Hu (1970) Potential photosynthesis and crop productivity; *Annals of the Association of American Geographers* 60, 92-101.

DROZDOV, A.V. (1971) The productivity of zonal terrestrial plant communities and the moisture and heat parameters of an area; *Soviet Geography* XII(1), 54-60. (Original in *Obshchiye teoreticheskiya problemy biologicheskoy produktivnosti*, Leningrad: Nauka, 1969, 192 pp.).

ELTON, C. (1927) *Animal Ecology*; (Macmillan & Co. New York), 207 pp.

FORMOZOV, A.N. (1964) Snow cover as an integral factor of the environment and its importance in the ecology of mammals and birds; *Materials for fauna and flora of the U.S.S.R.*, New Series, Zoology 5 (XX), 1-152 (translated by W. Prychodko and W.O. Pruitt Jr., Occasional Publication no. 1, Boreal Institute, University of Alberta, Edmonton, 1964, second printing 1969, 44 pp.).

GAASTRA, P. (1959) Photosynthesis of crop plants as influenced by light, carbon dioxide, temperature and stomatal diffusion resistance; *Mededelingen van de Landbouwhogeschool te Wageningen* 59, 1-68.

GAASTRA, P. (1963) Climatic control of photosynthesis and respiration; In L. EVANS (Ed.), *Environmental Control of Plant Growth*, (Academic Press, New York), 449 pp. (Specific reference pp. 113-40).

GERASIMOV, I.P. (1961) The moisture and heat factors of soil formation; *Soviet Geography* II(5) 3-12. (Papers of the Water-Heat Balance Symposium, Third Congress of the Geographical Society of the U.S.S.R.).

GRIGOR'YEV, A.Z. (1961) The heat and moisture regime and geographic zonality; *Soviet Geography* II(7), 3-16. (Papers of the Water-Heat Balance Symposium, Third Congress of the Geographical Society of the U.S.S.R.).

HARE, F.K. (1950) Climate and zonal divisions of the Boreal Forest formation in eastern Canada; *Geographical Review* 40, 615-35.

HARE, F.K. (1954) The boreal conifer zone; *Geographical Studies* 1, 4-18.

HARE, F.K. (1970) The tundra climate; *Transactions of the Royal Society of Canada* VIII, 393-99.

HARE, F.K. (1971) Snow-cover problems near the arctic tree-line of North America; *Reports of the Kevo Subarctic Station, University of Turku*, 8, 31-40.

HARE, F.K. and HAY, J.E. (1971) Anomalies in the large-scale water balance of northern North America; *Canadian Geographer* XV, 79-94.

HARE, F.K. and RITCHIE, J.C. (1972) The Boreal bioclimates; *Geographical Review* 62, 333-365.

HAY, J.E. (1970) *Aspects of the heat and moisture balance of Canada*; (Ph.D. thesis, University of London), 2 vols.

LETTAU, H. (1969) Evapotranspiration climatonomy: I. A new approach to numerical prediction of monthly evapotranspiration; *Monthly Weather Review* 97, 691-9.

LETTAU, H., and LETTAU, K. (1969) Shortwave radiation climatonomy; *Tellus* 21, 208-22.

LIETH, H. (1963) The role of vegetation in the carbon dioxide content of the atmosphere; *Journal of Geophysical Research* 68, 3887-98.

LINDEMAN, R.L. (1942) The trophic-dynamic aspect of ecology; *Ecology* 23, 399-418.

L'VOVICH, M.I. (1962) cited by R.G. BARRY, 1969: The world hydrological cycle, in *Water, Earth and Man*, R.J. Chorley (Ed.) (Methuen, London) 588 pp. (Specific reference pp. 12-29, especially figure 1.1.8.) (See also M.I. L'VOVICH, 1962: The water balance and its zonal characteristics; *Soviet Geography* III, 37-50.) (Original in *Izvestiya Akademii Nauk SSSR*, seriya geograficheskaya, 1962, p. 3-12).

McCORMICK, R.A. and LUDWIG, J. H. (1967) Climatic modification by atmospheric aerosols; *Science* 156, 1358.

MILLER, D.H. (1965) The heat and water budget of the earth's surface; *Advances in Geophysics* 11, 175-302.

MONIN, A.S. (1964) About heat-balance climatology; *Soviet Geography* V, 3-18. (Original precedes Budyko, 1964, q.v.).

MONTEITH, J.L. (1966) The photosynthesis and transpiration of crops; *Experimental Agriculture* 2, 1-14.

PALTRIDGE, G.W. (1970) A model of a growing pasture; *Agricultural Meteorology* 7, 93-130.

PETERSON, J.T. and BRYSON, R.A. (1968) Atmospheric aerosols: Increased concentrations during the last decade; *Science* 1962, 120-1.

PRIESTLEY, C.H.B. (1959) *Turbulent Transfer in the Lower Atmosphere*; (University of Chicago Press), 130 pp.

PRUITT, W.O., JR. (1970) Some ecological aspects of snow; In *Ecology of the subarctic regions*, (UNESCO, Paris), 83-99.

SELLERS, W.D. (1965) *Physical Climatology*; (University of Chicago Press), 272 pp. (Specific reference on p. 90).

TANSLEY, A.D. (1935) The use and abuse of vegetational concepts and terms; *Ecology* 16, 284-307.

THIENEMANN, A. (1918) Lebensgemeinschaft und Lebensraum; *Naturw. Wochenschrift*, N.F. 17, 282-90 and 297-303.

THIENEMANN, A. (1926) Der Nahrungskreislauf im Wasser; *Verh. Deutsch Zool. Ges.* 31, 29-79.

THIENEMANN, A. (1939) Grundzuge einen allgemeinen Oekologie; *Arch. Hydrobiol.* 35, 267-85.

THORNTHWAITE, C.W. (1948) An approach toward a rational classification of climate; *Geographical Review* 38, 55-94.

9 · Natural hazards research*
GILBERT F. WHITE

To a remarkable degree during the 1960's, geographers turned away from certain environment problems at the same time that colleagues in neighboring fields discovered those issues. This cluster of problems relates to the relationship between man and his natural environment, with particular reference to the kinds of transactions into which man enters with biological and physical systems, and to the capacity of the earth to support him in the face of growing population and of expanding technological alteration of landscape. In their self-conscious concern for developing the theoretical lineaments of a discipline, geographers tended to overlook those problems with which they, by tradition, had been concerned and which do not fall readily into allotted provinces of other scientific enterprises.

By neglecting the theory of man-environment relationships and its applications to public policy, the geographer loses an opportunity to apply his knowledge, skills, and insights to fundamental questions of the survival and quality of human life. He also fails to sharpen and advance theoretical thinking by testing it in a challenging arena of action. Any critical examination of man's activities as a dominant species in an ecosystem draws upon and invites refreshing appraisal by workers in other fields.

This argument is demonstrated by the line of natural hazard research as it has taken shape over the past fifteen years. It is presented here as an instance in which pursuit of a public policy issue led to a simple research paradigm and a model of decision-making dealing with how man copes with risk and uncertainty in the occurrence of natural events. The approach was refined and extended in a variety of situations, served to stimulate new methods of analyzing other geographical problems, and fostered a few changes in methods of environmental management by national and international agencies.

* The author is indebted to Ian Burton and Robert W. Kates for comments on an earlier draft.

The study and policy activities related in this direction of research represent an attempt to deepen understanding of the decision-making process accounting for particular human activities at particular places and times. The research seeks application of new techniques to one of the old and recurring traditions of geographical enterprise – the ecology of human choice. The results are slim yet promising. The experience may point more to errors to be avoided than to procedures to be emulated. However, the approach deserves appraisal as a possibly fruitful way of orienting new research and teaching of an old problem.

Application of this model and paradigm does not require any drastic changes in institutionalized teaching and research. Nor does it claim to establish a new sector of geographical inquiry. Rather, it offers one device for bridging some of the divergent lines of current investigation.

The problem

How does man adjust to risk and uncertainty in natural systems, and what does understanding of that process imply for public policy? This problem, raised initially with respect to one uncertain and hazardous parameter of a geophysical system – floods – in one country – the United States – provides a central theme for investigation on a global scale the whole range of uncertain and risky events in nature.

Genesis of the research

Definition of the problem had its genesis in observation of the results of a massive national effort in the United States to deal with the rising toll of flood losses. In 1927 the Corps of Engineers was authorized to conduct a series of comprehensive investigations to find means of managing the river basins of the United States for purposes of irrigation, navigation, flood control, and hydroelectric power. The legislative authorization called for the presentation to the Congress (the final decision-making body with respect to new construction projects on inter-state streams and tributaries thereof) of plans specifiying the needs of each area, the types of engineering construction work which could be undertaken, and projects proposed for Federal or State investment, giving the estimated cost and benefit. In the years following 1933 the so-called "308 reports" submitted to the Congress contained explicit benefit-cost analysis of possible construction projects.

In theory, to present a benefit-cost appraisal of a proposed project for a river basin required an analysis of the possible actions which man could take in managing the water and associated land resources of the area, and it also called for a systematic canvass of what, from the standpoint of society, would be the flows of social gains and losses to whomsoever they might accrue arising from any one of those interventions in the ecosystem. This was a monumental and presumptuous task.

Even in his most naive periods of technological mastery, man could not expect to understand the full set of consequences of any major interventions such as the channelization of the lower Mississippi River or the construction of a dam on the Upper Ohio or the building of a system of levees along the Sacramento. The investigator could make educated and hopefully intelligent guesses as to certain outcomes, e.g., alteration of stream regimen. He could not hope to identify all possible consequences. Measuring them would be still more difficult. Moreover, to complete a genuinely competent appraisal of possible lines of action would require canvass of the full range of possible activities which might be undertaken. A proposed dam then could be compared with other steps such as a levee, upstream management of vegetation, or downstream management of the flood plain. Yet, the practical engineering and administrative imperative was to go ahead with such investigations, using the best knowledge then available and applying an elementary kind of economic analysis in order to show for those items which could be readily quantified an estimate of prospective benefits and costs.

Thus, a program of planning took shape which was to have major consequences for resources planning and scientific work in other parts of the world as well as in the United States. Benefit-cost analysis of water projects in the United States became the most sophisticated piece of social impact investigation for several decades. There were more careful and detailed methodologies for computing water benefits and costs than for any other type of public investment. The procedures as first developed by the Corps of Engineers were later revised and embodied in rules and regulations issued by the Bureau of the Budget and approved by the Congress in two separate stages, and were the basis for extensive literature of economic analysis. The analysis was of a normative sort: it was designed to suggest ways by which estimates could be made as to the most effective investments to achieve specified public aims. Almost no time was given to finding out what in fact resulted from such investment. It was assumed that what was proposed— as, for example, the reduction of flood losses or the increase in waterway traffic — would in fact be realized if only the proper combination of technical means, discount rates, and time horizons could be found.

The 308 reports found their way into concrete action in a remarkably short period of time because they first appeared in the midst of the great economic depression of the 1930's and provided individual projects which could be used in mounting public works programs intended to relieve unemployment and stir economic recovery in the nation. The Tennessee Valley Authority was established with the intent, soon discarded, of using part of the Corps of Engineers 308 plan for that area. Large projects such as Grand Coulee Dam and the reservoirs in the Upper Ohio were authorized in the interest of revising a depressed economy.

Geographers early took an interest in this new line of planning but their more lively efforts either proved abortive or dwindled over a long period of time. They were active in the National Resources Planning Board — the first Federal agency in the United States to attempt to draw together the plans of independent state and national agencies into single, comprehensive river basin plans — and they joined in analysis of area economic and employment problems.

This interest stirred an investigation of the range of alternatives with respect to flood loss reduction (White 1942). It also stimulated the first comprehensive attempt to anticipate the full social impacts of a large impoundment. The impact study was carried out by the Bureau of Reclamation and associated agencies on the effects of the Grand Coulee Dam on the Columbia Basin (U.S. 1941). The latter work under the leadership of Harlan H. Barrows was not only a pioneer piece of interdisciplinary research, but defined in broad outlines and with notable gaps the problem of ecological impact which, while studied with considerable care for Grand Coulee, was not to be investigated again with similar energy or breadth until the late 1960's.

In 1936, following a series of disastrous floods affecting urban areas in the Mississippi system, the Congress authorized a national flood control policy which declared it to be the intent of the Federal government to contribute to the cost of flood control works wherever the anticipated benefits from such works would exceed the anticipated cost. In 1938 a supplemental act provided that where reservoirs were selected as a means of flood control no local contribution should be required to the cost of projects inasmuch as the allocation of benefits among the several state beneficiaries was so complicated that it seemed best to charge it all on the Federal account.

Twenty years passed, more than five billion dollars were expended on new Federal flood control works, and in 1956 a geographic investigation was begun of what had happened in the urban flood plains of the nation as a result of the investments during the two intervening decades. That investigation was to be followed by more thoughtful and searching studies through which ran the common thread of a relatively consistent research paradigm.

Research paradigm

In carrying out the 1956 appraisal of changes in land use in selected flood plains following the Flood Control Act of 1936, the geographic research group asked the following questions:

1. What is the nature of the physical hazard involved in extreme fluctuations in stream flow?
2. What types of adjustments has man made to those fluctuations?
3. What is the total range of possible adjustments which man theoretically could make to those fluctuations?
4. What accounts for the differences in adoption of adjustments from place to place and time to time?
5. What would be the effect of changing the public policy insofar as it constitutes a social guide to the conditions in which individuals or groups choose among the possible adjustments?

These questions were addressed to seven sites, chosen to give a diversity of conditions of floods, urban land use, and flood loss abatement measure (White, et al. 1958). A review also was made of the record for flood control expenditures and flood damages for the nation as a whole.

Adjustments were classified in three groups as shown in Table 9.1. From that view any human response to an extreme event in a natural system had the effect of (a) modifying the cause, (b) modifying the losses, or (c) distributing the losses.

A number of conclusions emerging from the field studies had an unsettling effect upon those who were responsible for Federal flood control programs, and triggered new investigations to probe unresolved questions. In brief, it was found that while flood-control expenditures had multiplied, the level of flood damages had risen, and that the national purpose of reducing the toll of flood losses by building flood-control projects had not been realized. Parts of valley bottoms were protected from floods, but increasing encroachment on the flood plain increased the damage potential from a smaller flood. One part of a city was protected by a levee, but new urban growth took place outside the levee. Works which controlled flood with a recurrence interval of 500 years were certain to fail with catastrophic consequences when the 1,000 year flood took place. The findings also indicated that because of the Federal government's concentration upon flood-control works and upstream water-

Table 9.1. Types of adjustments to floods

Modifying the Cause	Modifying the Loss	Distributing the Loss
Upstream land treatment	Flood protection works	Bearing the loss
	Dams	Public relief
	Levees	
	Channelization	Insurance
	Emergency measures	
	Flood warning	
Flood Proofing	Evacuation	
	Structural changes in buildings	
	Land elevation	

management activities to the exclusion of other obvious but relatively unpracticed types of adjustments, the situation was becoming progressively worse and showed no promise of being improved by a continuation of the prevailing policies. It was recognized that a rising flood toll might be beneficial if accompanied by larger benefits from flood plain use. However, the increased losses were contrary to the public expectation.

At that stage the study had (1) demonstrated that geographic research could have a direct bearing upon the formation of public policy in one country; and (2) posed a set of problems requiring further investigation if satisfactory policy readjustment was to be obtained. These problems centered on how to account for the differential behavior of individuals and groups in dealing with flood problems from one place to another. It had been shown that people did not behave as it had been expected that they would when the benefit-cost ratios for several thousand flood control projects had been drawn up. It was not equally clear why people had chosen the particular solutions they did and, therefore, what sorts of changes in public action would lead to genuine improvement in the character of their choices over a period of time. The effort to deal with this problem satisfactorily demanded further inquiry.

Models of decision-making

In all of the benefit-cost analysis and in the earlier work on changes in flood plain use, it was postulated that the choice made by people living on flood plains was essentially economic optimization. This was in the tradition of economic analysis, and conformed to the normative judgements on which the projects had been initiated. In essence, it assumed that individuals living in places of hazard would have relatively complete knowledge of the hazard and its occurrence, would be aware in some degree of the consequences, and would seek to make those adjustments which would represent an optimal resolution of the costs and benefits from each of the adjustments open to them. The ideal of the completely optimizing man was viewed as one rarely achieved in action, but as the framework within which a modified model, namely a model of subjective expected utility, might be explanatory. The subjective utility model held that man would seek to optimize but that his judgement would be based on incomplete knowledge and upon his subjective view of the possible consequences. It would be expected that if the view people had of the expected effects of using a particular piece of flood plain could be ascertained, it would be possible to judge their probable response by selecting those solutions which would give them the maximum net utility.

Neither the optimizing model nor the subjective utility model seemed to explain much of the behavior observed in the study areas. For example, it was found that although people seemed to recognize distinct differences in hazard from one part of the flood plain to another, they did not readily translate that recognition into differentials in assigned valuation of the property. People oftentimes returned to the use of land which had been severely damaged by floods being aware of the consequences of a recurrence and facing probable disaster of either a personal or financial character from such recurrence. Adequately to describe behavior for purposes of predicting responses to changes in public policy required the use of some other kind of model. Experimentation began with other possibilities. The obvious direction in which to move was the model of bounded rationality as described by Simon and others (Simon 1959). It was proposed in a general sense for a variety of resource management decisions, and was developed in a more rigorous fashion by Kates in his study of Lafollette, Tennessee (Kates 1962). In examining the behavior and expressed perceptions of residents of a flood plain in the Tennessee Valley, Kates attempted to find out how people perceive the hazard, how they perceive the range of adjustments open to them, and what factors accounted for differences in their perceptions. This required measurement of clearly economic gains and losses as perceived by them, but also consideration of a number of other factors such as the information available to the individual, his personal experience, and the physical nature of the event.

In the following years additional efforts were made to refine a model of bounded rationality, the most recent being that developed by Kates in connection with the collaborative research on natural hazards (1970). It will be noted from a simplified version as presented in Fig. 9.1 that a resource-management decision may be

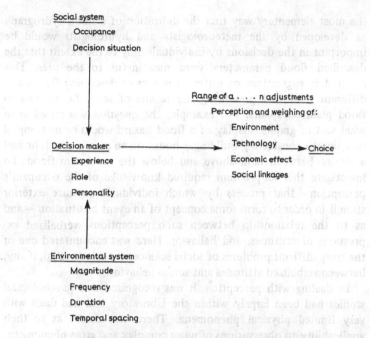

Figure 9.1. A Rough Model of Decision

hypothesized to involve the interaction of human systems and physical systems in terms of adjustment to a particular hazard. The interaction is represented as a choice-searching process as affected by personality, information, decision situation and managerial role. The result is a much more complicated model of how people make their choices in dealing with uncertainty and risk in the environment, one that did not lend itself as readily to careful field investigation, but that promised more revealing explanations of individual and group behavior.

The role of perception

As it was recognized that the judgment of the resource manager could be more important than the judgment of the scientific observer, other types of investigations were stimulated. It was clear in the most elementary way that the definition of a flood hydrograph as developed by the meteorologists and hydrologists would be important in the decisions by individuals only to the extent that the described flood parameters were meaningful to the user. This resulted in suggestions of different modes of describing floods and different modes of presenting the results of scientific analysis to flood plain occupants. For example, the question was raised as to what sort of graphic display of a flood hazard would be meaningful to a person considering building a house on an area in which he had a choice between land above and below the maximum flood. To investigate this subproblem required knowledge of the occupant's perception — that process by which individuals organize exterior stimuli in order to form some concept of an event or situation — and as to the relationship between such perceptions, verbalized expressions of attitudes, and behavior. Here was encountered one of the truly difficult problems of social science: the relationship, if any, between verbalized attitudes and actual behavior.

In dealing with perception, it was recognized that psychological studies had been largely within the laboratory and had dealt with very limited physical phenomena. There was doubt as to their applicability to observations of more complex and gross phenomena. Hewitt investigated the theoretical ground for expressing extreme events in probabilistic terms (1969). At this point the interest of geographers in problems of perception and attitude formation converged with those of psychologists, sociologists, city planners and architects who also were trying to specify perception and its implications (Burton and Kates 1964). Out of the concern for perception of floods came the first AAG symposium on problems of perception (Lowenthal 1967), a series of investigations dealing with perception of differing facets of the environment such as drought, recreational water, reservoirs, water supply alternatives, water re-cycling, and the like (Saarinen 1966; Baumann 1969), and Saarinen's geographic review of the perception literature (Saarinen 1969).

Out of the concern for attitudes developed a joint seminar between sociologists and geographers on problems of attitude formation, a joint investigation of attitudes toward water (White 1966), and a number of investigations bearing upon decision-making and public participation in such decisions (MacIver 1970; Johnson 1971).

Interdisciplinary cooperation

Much of the research could not have taken place without strong cooperation with workers in other disciplines. Engineers were essential to appraisal of the effects of physical structures, and they often were the key professional group in applying geographic findings. Part of the field investigations were supported by the Tennessee Valley Authority's division of Local Flood Relations, and its engineering personnel were a chronic source of critical encouragement. Members of the Corps of Engineers engineering staff participated in a few studies (Cook and White 1962), and the agency later invited and used the results of appraisals of operating experience in flood plain management. Hydrologists from the U.S. Geological Survey shared in the design and assessment of flood plain mapping and its presentation to public agencies.

Wherever an urban area was studied there usually was collaboration with responsible city planning officials. Their critiques were illuminating, and in some cases produced interesting new ventures such as the combination of geographic planning, and engineering skills in devising an urban redevelopment scheme for Waterloo, Iowa. However, it was necessary for geographic investigators to resist the temptation to become heavily involved in consulting activities. The pressures were to give time to applying the meager research findings rather than to expanding them.

Economists were drawn into the investigation, and contributed fresh insights into the process of optimal decisions. Unconventional views of flood losses as nature's rental for flood plains were developed by two of them (Renshaw 1961; Krutilla 1966). A refined method of assessing losses and benefits from land-use regulation was devised by Lind (1966). A more rigorous analysis of the economics of natural hazards was carried out by Russell (1970). An investigation of rural use of flood plains in the early 1960's was supported by the Agricultural Research Service of the Department of Agriculture, and brought geographers into working relations with agricultural economists (Burton 1962). However, it did not yield the anticipated refinements in economic aspects of flood hazard, and that sector of study awaits more intensive investigation.

Psychologists were drawn into examination of the personality traits affecting resource decisions. A simple sentence-completion test was devised by Sims (Natural Hazards Working Paper No. 16, 1970) and he collaborated in using thematic apperception tests (Saarinen and Sims 1969). A psychologist joined the research staff at the Department of Geography of the University of Toronto (Schiff 1971), and Kates collaborated with a psychology colleague in editing a review of hazard experience (Kates and Wohlwill 1966).

Interdisciplinary research in which workers in several fields genuinely interact is far more difficult to carry out than is research which draws from other fields at the pleasure of the investigator. In the latter case it is a readily manageable, sometimes gayly ostentatious, sometimes humbling exercise but always in he command of the investigator. When it is required by common commitment to solution of a problem, collaboration is not easily abandoned without personal hurt as well as cost to the whole enterprise.

Applications to public policy

It may help to briefly note a number of applications of this research paradigm and these models of decision-making to specific public policy issues. The interest in each case is two-fold: (1) what was its use in forming public policy and (2) what feedbacks, if any, did it have upon geographic theory?

At the outset of the flood plain occupance studies in 1957 an attempt was made to enlist the collaboration of people who were directly concerned in drawing up and carrying out such plans, and it was recognized that a principal alternative to the construction of flood control works was the regulation of land use. A representative of the Corps of Engineers took a year for a study leave to work with the Chicago geographic group and to produce a critical appraisal of experience with flood plain regulations (Murphy 1958). This led to tentative hypotheses as to community response to regulations and it also stimulated a legal investigation of the constitutional and statutory grounds for such regulation (Dunham 1959). The latter became the standard legal work on the subject.

Discussions of how people perceive the range of adjustments contributed to the establishment by the Corps of Engineers of a system of "flood plain information reports" which since have become operating practice. Early appraisals of providing flood plain information to residents of such areas showed the importance of individual perception in contrast to that of the scientific observer (Roder 1961), and fostered detailed experiments with modes of mapping sponsored by the U.S. Geological Survey (Sheaffer 1964). The Chicago metropolitan area became the first metropolitan area in the world to be completely mapped in terms of flood hazard. In the course of promoting and carrying out flood plain mapping through the Northeastern Illinois Metropolitan Planning Commission, further inquiries were made into the decision process. It became apparent that merely publishing the maps would be unlikely to have any significant effect upon decisions made by individuals or public agencies whereas if specific and favorable situations could be found in which the maps would be made available, the decision making might be changed. Thus, Sheaffer arranged for the organized group of land appraisers to make systematic use of the maps so that they in turn could attach a judgement about flood hazard to each land value assessment submitted to financial and mortgage agencies in connection with the purchase of buildings or property.

In addition, Sheaffer carried the first academic study of the possibilities of flood proofing, working jointly with personnel of the Tennessee Valley Authority (Sheaffer 1960). In time, that experience was the groundwork for preparing for the Corps of Engineers the preliminary manual of procedures for flood proofing for use of engineers, architects, and other technicians concerned with those alternative adjustments to floods (Sheaffer 1967). Geographers joined in the studies of receptivity that lay the foundation for the first fully operative national program of flood insurance in the United States (Czamanske 1967). They also helped sharpen the method of estimating flood losses (Kates 1965).

The eventual upshot of these investigations and their application in sample areas was the formulation of a new Federal flood policy under a task force in which geographers had a hand, established by the Executive Office of the President (U.S., 1966). The new policy involved basic changes in approaches and collaboration among nineteen different agencies, outlined a comprehensive effort by all interested agencies to deal with flood loss management, and inspired new lines of research and of data collection on their part

Table 9.2. Action recommended by the Task Force on Federal Flood Control Policy

---

To improve basic knowledge about flood hazard

The immediate listing of all urban areas with flood problems to alert the responsible agencies.

Preparation on maps and aerial photographs by the U.S. Geological Survey of reconnaissance delimitation of hazard areas.

More floodplain information reports from the Corps of Engineers and Tennessee Valley Authority

Agreement by federal agencies on a set of techniques to be used in determining flood frequencies.

A national program by the Corps of Engineers and Department of Agriculture for collecting more useful data on flood damages, using decennial appraisals, continuing records on sample reaches and special surveys after unusual floods.

Research by Department of Housing and Urban Development and USDA to gain greater knowledge on problems of floodplain occupance and on urban hydrology under the U.S. Geological Survey and HUD.

To coordinate and plan new developments on the floodplain

Specification by the Water Resources Council of criteria for regulation of floodplains and for treatment of floodplain problems.

Steps to assure that state and local planning would take proper and consistent account of flood hazard in:

Federal mortgage insurance (Federal Housing Authority and Veterans Administration)

Comprehensive local planning (HUD)

Urban transport planning (Bureau of Public Roads)

Recreational open space and development planning (Bureau of Outdoor Recreation)

Urban open space acquisition (HUD)

Urban renewal (Urban Renewal Administration and Corps of Engineers)

Sewer and water facilities (HUD, USDA, Department of Health, Education and Welfare, and Ecomomic Development Administration)

Consideration by Office of Emergency Planning, Small Business Administration, and Treasury Department of relocation and flood-proofing in rebuilding flooded areas.

A directive to all federal agencies to consider flood hazard in locating new facilities.

Table 9.2. (Cont.)

To provide technical services to managers of floodplain property

    Collection and dissemination of information by Corps of Engineers in collaboration with USDA and HUD on alternative methods of reducing flood losses.

    An improved system of flood forecasting under Environmental Sciences Service Adminstration.

To move toward a practical national program for flood insurance

    A brief study by HUD on the feasibility of insurance.

To adjust federal flood control policy to sound criteria and changing needs

    Broadened survey authorizations for Corps of Engineers and USDA.

    Provision by the Congress for more suitable cost sharing by state and local groups.

    Reporting of flood control benefits to distinguish protection of existing improvements from development of new property.

    Authorization by the Congress to include land acquisition as part of flood control plans.

    Authorization by the Congress of broadened authority to make loans to local interests for their contributions.

(see Table 9.2). An Executive Order (Number 11296) at the same time required all government agencies responsible, directly or indirectly, for locating new buildings on flood plains to take account of flood hazard in the location decision.

It would be a mistake to suggest that the resulting policy has been fully or effectively translated into action in all responsible agencies. Any basic change in bureaucratic outlook is slow at best. Yet, part of the geographic view of flood plain adjustments had been adopted within four years. A geographer had been appointed to head the flood plain study section of the Corps of Engineers. While reasonable progress was made by most units of government, several agencies dragged their heels against revisions in their procedures, as when the Department of Agriculture committed itself to building flow regulation and land treatment structures to the virtual exclusion of other types of adjustments (White 1970a).

In a number of states such as Iowa, Nebraska, and Ohio, geographers played a part in instigating and carrying out state efforts to apply the same comprehensive approach to flood problems in their respective areas. Under the leadership of geographers the Center for Urban Studies at the University of Chicago initiated several appraisals of experience with flood plain management which assisted in revision of Federal operating policies.

Certain of the activities recommended by the Task Force found interested response in other countries. Thus, the preparation of maps of flood prone areas was undertaken in France under the sponsorship of the Ministere de l'Equipement et du Logement (France, n.d.; 1968). Studies of flood problems were sponsored by government agencies in Canada (Sewell 1965).

At the international level, the Department of Economic and Social Affairs of the United Nations joined with the Ministry of Reclamation and Water Management and other agencies in the USSR in sponsoring in 1969 a Seminar on Methods of Flood Loss Management. The Seminar brought together specialists, primarily engineers, from 28 developing countries and a number of consultants, including one geographer each from Canada, Japan, the United Kingdom, and the United States (White 1970b). They gave careful thought to the approaches initiated in the United States, and in some countries the effects are now observable in national study activity. The United Nations report on the Seminar gives the Seminar findings, points out the implications of geographic research, and suggests new flood loss reduction policies for informal guidance of officials coping with flood losses in developing countries.

The approach which seemed to be yielding results in the realm of flood losses was given application in several other sectors. The Office of Science and Technology established a special Task Force on Earthquake Problems which was patterned after the experience of the Task Force on Federal Flood Control Policy and which benefitted from the geographic contributions to the National Academy of Sciences report on the Alaska earthquake. Russell, Kates, and Arey pursued the problem of optimization in dealing with drought hazard as related to municipal water supply following the New England drought of 1965 (1970).

Collaborative studies on natural hazard

The various threads of inquiry were drawn together again in 1968 by a collaborative investigation of natural hazards supported by the National Science Foundation. Burton of Toronto and Kates of Clark joined with the author in examining the experience with a large array of hazards – drought, earthquake, flood, frost, landslide, hurricane, snow, tornado, and volcano – in a variety of settings (Burton et al. 1968).

Scientific as well as public policy response to these activities was sufficiently promising so that the International Geographical Union Commission on Man and Environment decided in 1969 to adopt as one of its two principal thrusts in the succeeding three-year period a program for international collaboration in the study of problems of environmental hazards. These joint investigations now comprise a number of comparative field observations and national studies as outlined in Table 9.3.

The selection of study areas and collaborators was in many instances fortuitous: areas were chosen in terms of the inherent interest of the occupance and environmental problems but with a practical eye to the availability of competent personnel to carry out the work. It is hoped that from them will come a more rigorous and searching testing of a number of the hypotheses that slowly have emerged over the years since the first office analysis was made of the range of adjustments to floods. Some of the early findings no doubt will be reversed or discarded. The principal hypotheses are now under examination by the collaborators at Clark, Colorado and Toronto.

Perhaps the most fundamental of those hypotheses is that rational explanations can be found for the persistence of human occupance in areas of high hazard by examining the perception of the occupants of such areas and searching out their views of the alternative adjustments and the likely consequences of adopting any one of those opportunities.

In general, we suspect that there are three major types of response to natural hazards. Tentatively, we characterize these as follows: (1) Folk, or pre-industrial response, involving a wide range of adjustments requiring more modifications in behavior and harmony with nature than control of nature and being essentially

Table 9.3. Field studies as part of collaborative research on natural hazards, 1971

Comparative Observations of Small Areas

Coastal Erosion	Scotland
	United States
	Wales
Drought	Australia
	Brazil
	Mexico
	Tanzania
	United States
Earthquake	Peru
	Sicily
	United States
Flood	Canada
	Ceylon
	France
	India
	Japan
	United Kingdom
	United States
Frost	United States
Hurricane	Pakistan
	United States
Landslide	Japan
	United States
Snow	United States

National Studies of One Hazard

Drought	Australia
	Tanzania
Hurricane	East Pakistan
	United States
Flood	Ceylon
Air Pollution	United Kingdom

---

flexible, easily abandoned, and low in capital requirements. (2) Modern technological, or industrial response involving a much more limited range of technological actions which tend to be inflexible, difficult to change, high in capital requirement, and to require interdependent social organization. (3) Comprehensive, or post-industrial, response combining features of both of the other types, and involving a larger range of adjustments, greater flexibility, and greater variety of capital and organizational requirements. We hypothesize that the United States currently is passing the peak of the modern technological type and is beginning to catch glimpses of the comprehensive type as it emerges here and elsewhere, but we do not suggest that there is a necessary sequence in the types of response.

It is also hypothesized that variations from place to place in hazard perception and estimation can be accounted for in considerable measure by a combination of factors embracing (1) certain physical characteristics of the hazard, (2) the recency and severity of personal experience with the hazard, (3) the situational characteristics of decisions regarding adjustments to the hazard, and (4) personality traits.

We have been inclined to try to describe choice of adjustment in terms of a perception model dealing with the individual manager's subjective recognition of the hazard, of the range of choice open to him, the availability of technology, the relative economic efficiency of the alternatives, and the likely linkages of his action with other people.

We further hypothesize that there are significant differences in the way in which these factors interact in relation to community action in contrast to individual action.

The alternatives approach

Another spinoff from the early flood plain investigations was application of the idea of range of alternative adjustment to other aspects of natural resources management. In elementary terms the alternatives approach in flood losses could be adopted to any other purposeful intervention in the environment. For example, in combatting stream pollution, the building of waste treatment works or of storage for diluting stream flow are only two of a much larger range of adjustments (Davis 1968). Alternatives would include such measures as controlling waste at the source, use of waste in agriculture, oxygenating streams, constructing special channels for waste transport, and the like.

This view was expressed in two reports from the National Academy of Sciences Committee on Water, with geographic participation (NAS 1966, 1968); that had a significant effect on national water policy in the late sixties. Attention to the full range of practicable adjustments converged with concern for systems analysis to produce water planning methods found in the North Atlantic Regional Water Study. In the North Atlantic study all interested Federal agencies and twelve states join in preparing river basin reports showing for each of three alternative aims (national economic efficiency, regional development, and environmental preservation) the range of possible activities, including non-structural devices, which might be undertaken to meet perceived needs.

The same approach was embodied in part of the High School Geography Project. Its unit on environmental study introduces the student to analysis of alternative ways of dealing with flood losses in an industrial area. The treatment there coincides with increasing emphasis in the social scene on consideration of the range of possible social action in contrast to dependence upon simple technological solutions.

Appraisal

It is too early to venture an appraisal of how influential this direction of natural hazards research has been upon either public policy or geographic thinking; nor are we well equipped to try. In the short run it clearly has been linked to changes in methods of managing water and associated land resources in one nation, and to a smaller degree in several others. What effect those changes will have in the long run is impossible to predict. As of 1971 they pointed to more searching examination of the range of choice available to man in coming to terms with his environment. Although the research has made only modest contributions to a theory of man-environment relations, it supported new efforts to specify the nature of environmental perception, to recognize the process of decision making for resource management, and to identify the landscape consequences of alternative public policies.

In essence, the activity was problem oriented and interdisciplinary. Such work is often tiresome and sometimes exhilarating. It requires research findings in a form highly intelligible to workers in other fields. It ignores conventional divisions of an academic field.

One lesson emerges from this history of investigation of a single environmental problem, using a rather unsophisticated research paradigm. It is that if environmental problems are pursued rigorously enough and with sufficient attention to likely contributions from other disciplines they may foster constructive alternations in public policy but at the same time may stimulate new research and refinement of research methodology to the benefit of geographic discipline. Both may serve to advance man's painful, faltering, and crucial struggle to find his harmonious place in the global systems of which he is a part.

References

BAUMANN, D.D. (1969) The recreational use of domestic water supply reservoirs: Perception and choice; *University of Chicago, Department of Geography Research Paper No. 121.*

BURTON, I. (1962) Types of agricultural occupance of flood plains in the United States; *University of Chicago, Department of Geography Research Paper No. 75.*

BURTON, I. (1965) Flood damage reduction in Canada; *Geophysical Bulletin* 7, 161-85.

BURTON, I. and KATES, R.W. (1964) Perception of natural hazards in resources management; *Natural Resources Journal* 3, 412-41.

BURTON, I., KATES, R.W. and SNEAD, R.E. (1969) The human ecology of coastal flood hazard in megalopolis; *University of Chicago, Department of Geography Research Paper No. 115.*

BURTON, I., KATES, R.W. and WHITE, GILBERT, F. (1968) The human ecology of extreme geophysical events; *Natural Hazard Research, Working Paper No. 1, Department of Geography, University of Toronto.*

COOK, H.L. and WHITE, GILBERT, F. (1962) Making wise use of flood plains; In *United Nations Conference on Applications of Science and Technology,* (Government Printing Office, Washington), Vol. 1, 343-59.

CZAMANSKE, D.V. (1967) *Receptivity to Flood Insurance;* (Master's Dissertation, University of Chicago).

DAVIS, R.K. (1968) *The Range of Choice in Water Management: A Study of Dissolved Oxygen in the Potomac Basin;* (Johns Hopkins Press, Baltimore).

DUNHAM, A. (1959) Flood control via the police power; *University of Pennsylvania Law Review* 107, 1098-132.

FRANCE, Ministère de l'Equipment et du Logement, *Inventaire des Zones Inondables*; (Paris: BCEOM, n.d.).

FRANCE, Ministère de l'Equipment et du Logement (1968) *Etats-Unis: Recherches Methodologiques sur la Rentabilite Economique des Mesures de la Controle des Crues a L'Etranger*; (Paris, BCEOM).

GODDARD, J.E. (1971) Flood plain management must be ecologically and economically sound; *Civil Engineering* September, 81-5.

HEWITT, K. (1969) Probabilistic approaches to discrete natural events: A review and theoretical discussion; *Natural Hazard Research Working Paper No. 8, Department of Geography, University of Toronto.*

JOHNSON, J.F. (1971) Renovated waste water; *University of Chicago, Department of Geography Research Paper No. 135.*

KATES, R.W. (1964) Variation in flood hazard perception: Implications for rational flood plain use; In *Spatial Organization of Land Uses: The Willamette Valley*, (Oregon State University, Corvallis).

KATES, R.W. (1962) Hazard and choice perception in flood plain management; *University of Chicago, Department of Geography Research Paper No. 78.*

KATES, R.W. (1965) Industrial flood losses: Damage estimation in the Lehigh Valley; *University of Chicago, Department of Geography Research Paper No. 98.*

KATES, R.W. (1971) Natural hazard in human ecological perspective: Hypotheses and models; *Natural Hazard Research Working Paper No. 14, Department of Geography, University of Toronto.*

KATES, R.W. and WOHLWILL, J.F. (Eds.) (1966) Man's response to the physical environment; *Journal of Social Issues* 22, 1-140.

KRUTILLA, J.V. (1966) An economic approach to coping with flood damage; *Water Resources Research* 2, 183-90.

LIND, R.C. (1966) *The Nature of Flood Control Benefits and the Economics of Flood Protection*; (Stanford University Institute for Mathematical Studies in the Social Sciences).

LOWENTHAL, D. (Ed.) (1967) Environmental perception and behavior; *University of Chicago, Department of Geography Research Paper No. 109.*

MacIVER, I. (1970) Urban water supply alternatives: Perception and choice in the Grand Basin, Ontario; *University of Chicago, Department of Geography Research Paper No. 126.*

MILLER, D.H. (1966) Cultural hydrology: A review; *Economic Geography* 42, 85-9.

MURPHY, F.C. (1958) Regulating flood plain development; *University of Chicago, Department of Geography Research Paper No. 56.*

NATIONAL ACADEMY OF SCIENCES COMMITTEE ON WATER (1966) *Alternatives in Water Management*; Publication 1408, (Washington, D.C.).

NATIONAL ACADEMY OF SCIENCES COMMITTEE ON WATER (1968) *Water and Choice in the Colorado Basin: An Example of Alternatives in Water Management*; Publication 1689, (Washington, D.C.).

NATIONAL HAZARD RESEARCH (1970) Suggestions for comparative field observations of natural hazards; *Natural Hazard Research Working Paper No. 16, Department of Geography, University of Toronto.*

RENSHAW, E.F. (1961) The relationship between flood losses and flood control benefits; In 'Papers on Flood Problems', *University of Chicago, Department of Geography Research Paper No. 70.*

RODER, W. (1961) Attitudes and knowledge in the Topeka flood plain; In 'Papers on Flood Problems', *University of Chicago, Department of Geography Research Paper No. 70.*

RUSSELL, C.S. (1970) Losses from natural hazards; *Journal of Land Economics*, Vol. 46, p. 38-

SAARINEN, T.F. (1969) Perception of environment; *Association of American Geographers, Commission on College Geography*, (Washington, D.C.).

SAARINEN, T.F. (1966) Perception of the drought hazard on the *American Geographers, Commission on College Geography*, (Washington, D.C.).

SCHIFF, M.R. (1971) Psychological factors relating to the adoption of adjustments for natural hazards in London, Ontario; *Paper presented to Association of American Geographers, Boston, April 1971.*

SEWELL, W.R.D. (1965) Water management and floods in the Fraser River Basin; *University of Chicago, Department of Geography Research Paper No. 100.*

SHEAFFER, J.R. (1960) Flood proofing: An element in a flood damage reduction program; *University of Chicago, Department of Geography Research Paper No. 65.*

SHEAFFER, J.R. (1964) Economic feasibility and use of flood maps; *Highway Research Record* 58, 44-6.

SHEAFFER, J.R. (1967) *Introduction to Flood Proofing: An outline of principles and methods*; (University of Chicago, Center for Urban Studies).

SIMS, J. and SAARINEN, T.F. (1969) Coping with environmental threat: Great Plains farmers and the sudden storm; *Annals of the Association of American Geographers* 59, 677-86.

UNITED STATES BUREAU OF RECLAMATION (1941) *Columbia Basin Joint Investigations: Character and Scope*; (Government Printing Office, Washington, D.C.).

UNITED STATES, Eighty-ninth Congress, Second Session (1966) *A Unified National Program for Managing Flood Losses*; House Document 465, (Washington, D.C.).

WHITE GILBERT, F. (1942) Human adjustment to floods; *University of Chicago, Department of Geography Research Paper No. 29*.

WHITE, GILBERT, F. (1964) Choice of adjustment to floods; *University of Chicago, Department of Geography Research Paper No. 93*.

WHITE, GILBERT, F. (1966) Formation and role of public attitudes; In Jarrett (Ed.), *Environment Quality in a Growing Economy*, (Johns Hopkins Press, Baltimore).

WHITE, GILBERT, F. (1966) Optimal flood damage management: Retrospect and prospect; In Kneese and Smith (Eds.), *Water Research*, (Johns Hopkins Press, Baltimore).

WHITE, GILBERT, F., CALEF, W.C., HUDSON, J.W., MAYER, H.M., SHEAFFER, J.R. and VOLK, D.J. (1958) Changes in urban occupance of flood plains in the United States; *University of Chicago, Department of Geography Research Paper No. 57*.

WHITE, GILBERT, F. (1969) *Strategies of American Water Management*; (University of Michigan Press, Ann Arbor).

WHITE, GILBERT, F. (1970a) Flood loss reduction: The integrated approach; *Journal of Soil and Water Conservation* 25, 172-6.

WHITE, GILBERT, F. (1970b) Recent developments in flood plain research; *Geographical Review* 60, 440-3.

PART IV

Temporal

10 · Forecasting alternative spatial, ecological and regional futures: problems and possibilities*

PETER HAGGETT

By the late 1960s powerful push and pull elements were forcing a small but increasing proportion of geographers to move their energies towards the forecasting arena. If we take the three main strands in geographic research we find broadly similar shifts occurring. Within the *spatial* tradition, locational analysis was running into something of an impasse. On the 'supply' side, geographic research was still largely concerned with tuning up descriptive and static models of existing urban and regional structures. On the 'demand' side, the needs of the community, from international to local levels, was for dynamic models of an interventionist kind. Specifically there was a need for calculations by geographers which would allow the spatial impact of alternative intervention strategies to be evaluated.

Within the second tradition, that of *ecological* analysis, similar pressures were discernable. Again ecological work, whether at the global level of population-food equations or at the local level of river or atoll ecosystems, was being dragged into the public arena. Under the advocacy of Commoner, the Ehrlichs, Earth Days, Royal Commissions, and the like, geographers were being forced to put static ecological models forward in a market-place that clamoured for the dynamic; the emphasis was on using such models to predict likely directions and rates of environmental change. For the third tradition, that of *regional* geography, the signs were less easy to read.

* This essay is a shortened version of an extended review of the use of forecasting models in geography forthcoming in *Progress in Geography* 7 (1973).

Within the field itself, regional geographers had stubbornly refused to be written off like Victorian antimacassars (to use John Paterson's term in introducing his revised *North America* (1970, iv.)). Instead, an increasing proportion of the raw material for expanding area studies became forward looking. No regional geography of the United Kingdom could now be written without taking into account the growing shelves of future-orientated studies — national plans, regional planning council reports, metropolitan projections, and the like.

Some notion of the ladder up which we might progress from the current stock of descriptive models to the newly desired ends of dynamic planning models was provided by Lowry (1965). He saw *descriptive, forecasting,* and *planning* models as three discrete stages in the evolution of an evolving sequence of model-building. By the end of the decade the signs were that each of the three intertwined strands of geography (Haggett 1972, 452-4) were twisting towards the second stage in a heliotropic attempt to move upwards towards the third. Here we look at the first of these strands.

The existing pattern

The "geography of the future" is an emotive phrase, conjuring up shades of H.G. Wells, Jules Verne, or George Orwell. As Ryan's (1971) survey showed, forecasting is intricately associated with images of horology, futurology and soothsaying. But the few geographers who have chosen to write about future spatial or ecological patterns choose to write in a more restrained vein. Thus Peter Hall's *London 2000* (Hall 1969 edn.) is a careful projection of the distribution of people, homes, jobs and travel in the metropolis and the problems of planning and organization they pose. Similarly Brian Berry's work on the future spatial patterns in the United States (Berry 1971) is for the most part a projection of trends already making themselves felt. Likewise those physical geographers who took part in the 1965 conference on Future Environments of North America (Fraser Darling and Milton 1966) deal in terms of the ecological implications of well-substantiated trends in land use.

It would of course be misleading to suggest that only geographers who had specifically included with term "future" in their writing were concerned with the period beyond the present. Many works in physical, human or regional geography contain important implications about the future — if only in terms of an assumption of continuity. It remains true however that geographers as a whole have been concerned with local regional or global yesterdays rather than tomorrows. Let us illustrate this by taking from the shelf all the geographical periodicals published in a single year in the 1960s, say 1965. If we now examine those papers which are concerned with particular landscapes or regions (rather than those dealing just with abstract models or techniques) then we find that most describe what we may term the "historical present" — that is they describe a situation using the latest available statistics, air photographs, field observations, maps, and so on. If we add together the time lag between the date at which this information was collected and its subsequent use by the investigator, and a second time lag between the completion of the work, we have a period which is rarely less than one year and more often four to five. In other words, even those studies which describe the present geography of an area are necessarily describing its recent past. But we must add to these studies those papers which are explicitly describing the past; the historical geographer describing the Ante Bellum South or the geomorphologist describing a post-glacial strandline. The resulting frequency distribution of papers in the 1965 journals displays a highly skewed form with its peak around 1960 and a long tail stretching back into past years. The distribution is sharply truncated at around 1964 with a very few outliers devoted to the geography of periods beyond 1965 itself.

If this view is a correct one, then what is its significance? Clearly to some extent the time lag is simply a product of the gap between data collection and analysis, and between analysis and publication. Where research is dependent on a decennial census we should expect some consequent periodicity in any cross-sectional studies. But this restriction is equally valid for other subjects — Econometrics or Demography — fields which are noted for their concern with forecasting and projecting future social and economic patterns. We must leave the historians of the subject to unravel the reasons for this diffidence within geography, but probably the additional complications imposed by adding the spatial constraint to that of time itself may be seen to have played a significant part.

Some initial questions

Before looking at the strategies by which we can build forward from
descriptive models towards planning models, through the intervening
medium of the forecasting model, we need to clear some important
hurdles. What do we mean by the term forecasting, and should
geographers be in the forecasting business anyway?

The terminology of forecasting

The language used by forecasters, be they biblical or biological, has
given rise to a series of difficulties. Jantsch (1967) in his review of
alternative forecasting methods for OECD provides a useful glossary
of terms. An important distinction is drawn between apodictic
(non-probabilistic) statements about the future on an absolute
confidence level *(predictions)*, and similar statements about the
future based on a stated confidence level *(forecasts)*. Where the
confidence level is low or unable to be defined, Jantsch prefers the
use of the term *anticipations*. Closely akin to anticipations are the
loosely-structures *surmises* of the many "surmising forums" like the
Hudson Institute, the Futuribles group in France, or SSRC's
'Committee on the Next Thirty Years'. These recently-founded
centres use the interaction of ideas in small groups to identify some
common ideas on the interwoven elements in alternative futures and
the critical boundary points which lead to one alternative rather
than the other.

A second important distinction relates to the starting point of
any statement about the future. Thus we can distinguish between an
explorative forecast which takes it's starting point as the present and
works forward into the future, and a *normative forecast* which first
assesses externally the future goals and needs and uses these to work
backwards towards the present. Some workers prefer to use the term
projections to describe the first case although this is often still
further restricted to the extrapolation of curves fitted to a given set
of past observations.

In principle, many of the forecasting methods can be extended
either forward *or* backward in time. Thus *simulation* models can be
built to replicate the observed behaviour of a system (with

computers, human players, or both) and allowed to run on into the future, or backwards into the past. Similarly, trend projections could be extrapolated backwards to reconstruct missing historical information. Ugly terms like "retrodiction" or "postdiction" have been proposed to describe this process (see review of use in historical geography by Prince (1971)).

Past forecasts of future geographies

Willingness to make a fool of oneself has been proposed (by one of its more distinguished members) as one qualification for membership of the General Systems Society; similar spurs tend to be won by geographic forecasters. The range of misunderstandings and miscalculations is wide, and the literature strewn with the spent cartridges of old projections. In the early eighteenth century Montesquieu had no doubts about the future level of world population: "After performing the most exact calculation possible in this sort of matter, I have found that there is scarcely one tenth as many people on the earth as in ancient times. What is surprising is that the population of the earth decreases every day, and if this continues, in another ten centuries the earth will be nothing but a desert" (*Lettres persanes,* 1721; cited by Jouvenel 1967, 13). Greater success has been achieved at the level of the single country, the United States, and Berry (1970, 22-3) finds that the surmises of Robert Vaughan in 1843 and H.G. Wells in 1902 were disturbingly accurate — in the general pattern of change, if not in its timing. A limited survey of past forecasts of future geographies shows a chastening mixture of insight and delusion.

In the light of our limited experience, what are the proprieties of geographers studying the future. A parallel debate has been running for some years amongst our historical colleagues with Heilbroner's *The future as history* (1959) and Plumb's *The death of the past* (1970) as extreme poles. The facts are that future geographies are already being written. by non-geographers as part of general forecasting scenario. Thus Kahn and Wiener's *The Year 2000* (1967) contains a region-by-region world picture of population and income levels at the end of the century. Over shorter forecast periods, geographers are being drawn into national, regional, and local projections as part of planning exercises (see, for example, Willatts

(1971) review of the involvement of geographers in the United Kingdom in this planning process).

In practice "to forecast or not to forecast?" generally centres on pragmatic cost grounds rather than academic niceties. If the cost of making a forecast error is very large, then this normally justifies the costs of setting up an elaborate forecasting machinery and *vice versa*. Where the penalties for making a wrong forecast or no forecast at all are nil, then the spur to forecasting is missing. Of course, the benefit-cost ratio is very crude and we may think of elaborate forecasting inquiries that augmented rather than reduced the errors likely from an informed guesstimate: as Colenutt (1970) has shown, the relationship between model complexity and accuracy is not a simple or linear one. The problem is made more difficult by the assymetric nature of the costs of forecast errors. In their pessimistic view of the future world food-population balance, Ehrlich and Ehrlich (1970) stress that the dangers of underestimating the future problems are far graver than those of overestimating. They lay stress therefore on possible future events which, although unlikely, will be catastrophic if the do occur. A number of important long-term forecasts have this characteristic lack of symmetry in the utility function of their errors.

Alternative forecasting strategies

If we agree that, despite its pitfalls, some form of forecasting is desirable or at any rate, inevitable, then how do we proceed? A review of the variety of procedures used in well established physical fields (e.g. meteorology) supplemented by the more recent experience of the social sciences (e.g. econometrics) reveals a surprisingly large range of procedures. The survey by Jantsch (1967) identified over one hundred methods in the field of social and economic forecasting alone.

We can reduce this range to six main families of methods. The first two are largely "black box" in approach since they rely on the repetition of patterns in time and space without attempting to build up causal chains. Conversely, the latter four involve (to a varying degree) the inclusion of other factors in the model other than the one being forecast. In line with conventional practice we term the variable being forecast the "predictand" or "dependent variable (Y)", and the variables used to aid the forecast as the "predicators" or "independent variables $(X_1, X_2, \ldots X_n)$".

Procedure 1. Forward extrapolation of apparent *trends* or *periodicities* in the predictand. Methods may range from filtering of trend curves using simple graphical techniques, (Gregg, et al 1964) through to very sophisticated spectral analysis (Granger, 1964) demanding long time series and computer back up.

Procedure 2. Analogue matching in which predictand record is searched to find closely similar cases. The prediction is based on the subsequent development of the analogue pattern, using either the most frequently recurring case, or the average of all subsequent developments.

Procedure 3. Statistical regression analysis and related techniques using time-lagged correlations between the predictand and one or more predictor variables. Although among the most widely-used procedures, this family of methods runs into severe theoretical problems related to multicollinearity and autocorrelation (see review by Colenutt 1970).

Procedure 4. Contingency tables in which the possible range of both the variable to be predicted and the variable or variables used as predictors are divided into a number of classes. The frequency of occurrence of each predictand class is compared to the prior occurrence of combinations of estimating variables.

Procedure 5. Complex simulation models in which the interactions of the predictand and estimator variables are built into a theoretical model which allows both positive and negative feedbacks. Informal examples are based on a forecaster's individual judgement when the integration is based on experience plus the application of current theoretical concepts. To an increasing extent, such integration is being replaced by numerical simulations (e.g. Monte Carlo models) using high-speed digital computers.

Procedure 6. Informal *scenario* or *Delphi* methods in which the views of a group of forecasters are used to build up a combined view of a complex situation that appears to be unamenable to a simulation model. This method is increasingly used to draw up long-term surmises over future ecological or socioeconomic patterns; and is usually supported by more formal forecasts of some elements in the situation using one or more of procedures (1) to (5).

Each of the above represents a family of techniques rather than a single method. Thus we can subdivide the various trend methods to include a sub-clan concerned with forecasting changes or rates of change in the predictand rather than its future level. Not all the inquiries are directed towards predicting values at a *specific* point in time. For example, in hydrological forecasting we may be interested simply in estimating the probable occurrence of an event of given magnitude (e.g. the size of the 100-year flood) rather than the exact timing of the event. Physical geographers dealing with events that may reasonably be regarded as recurrent are particularly interested in this set of problems (Wolman and Miller 1960).

Spatial stucture of forecasts

Some general cases

In terms of their geographic interest, the differences between the six groups of forecasting methods is less important than their spatial structure. Let us assume that we wish to predict the future value ($t +$ m) of a variable Y at a location i, and that we are making the forecast at time t. Let us further assume that we estimate Y in terms of a simple linear regression at X_t. We can now envisage three simple spatial situations:

Case I: Value for X_t are available for the same location as point i. In this *identical* case the predicting equation is simply

$$Y_i(t + m) = \alpha + \beta X_i(t) + \epsilon$$

where ϵ defines an error or disturbance term.

Case II: Values for X_t are not available for point i but are available for the aggregate of location (I) of which i forms a part. In this *hierarchic* case

$$Y_i(t + m) = \alpha + \beta X_I(t) + \epsilon$$

Case III: Values for X_t are not available for point i but are available for an alternative point (j). In this *contagious* case

$$Y_i(t + m) = \alpha + \beta X_j(t) + \epsilon$$

If we relax each of the restrictions in the simple model we can accommodate most of the six methods within these terms. For example, if we allow the predictor X to include the present and previous values of Y then we can accommodate all autoregressive and trend projection methods. Likewise if we expand the equations to allow a set of estimating $(X_1, X_2 \ldots X_n)$ and increase the number of locations to include points other than i and j we can absorb the analogue matching methods. If we restructure the equations to allow lagged feedbacks then we can also accommodate the more complex simulation methods.

Local trend projections

Most of the classical forecasting studies fall within the first spatial case. Thus national forecasts of future population levels and future economic activity tend to be conducted in terms of past sets of values for the nation itself. To be sure they will include estimates of flows into and out of the nation whether in terms of migration levels or foreign investment, but typically these are conducted in aggregated terms or as simple ratios rather than being spatially decomposed. Geographers tend therefore to be more interested in the second and third cases.

Hierarchic relations between regions

By hierarchic forecasts we desribe methods by which the levels or changes of a predictand variable in a local area (i) are forecast in terms of the levels or changes of either the predict and variable or other variables in a larger region (I). There may of course be a hierarchy of spatial levels separating i and I but here we restrict ourself to the simple two-stage case.

Typically if we compare the time series of a local area with those of the region of which it forms a part we see that the peaks and troughs are broadly comparable although they vary somewhat in detail. We can express the variation between the two series by regarding the series for the larger area as an *input* and the smaller areas as an *output*. Such inputs and outputs may be simply related through an input/output transformation. Thus a transformation may have an amplifying effect in which a small swing at the regional level is transformed into a major swing at the local level. We know for example that modest business-cycle fluctuations in the national economy may cause grave local repercussions in areas that have a heavy concentration of employment in certain industries (e.g. the steel industry). Although the reasons for this greater vulnerability to cycles are complex, we may broadly say that in slump conditions it is possible to postpone a decision to buy a new car, refrigerator, tractor (all needing steel), but we continue to need to buy food, pay taxes, or go to college. Thus small areas specializing in this latter group (say, a small university town) may show smaller fluctuations than the national average. In terms of our input-output transformation we would see this as a deviation-dampening effect.

We may wish to forecast the regional in terms of the local for a number of reasons. First, the regional series may be easier to predict since it is less susceptible to sudden irregular events (e.g. the closure of a single factory) that have a distorting effect on the local series. In particular the long-term or secular trend may be more stable. Second, we may be able to check the validity of the series we are interested in against trends in other series. Thus we could check our figures for employment against those of production, or trade, or investment, which are not likely to be available at the local level. They provide a check on whether the employment trends match up with our forecasts for these other related variables. It is clear that the form of the relationship will depend on the boundaries of the region. There are a large set of possible I's which include i within their limits and they may be very different errors associated with the choice of region.

Bellwethers and contagious relations

The third class of interpolated forecasts includes methods by which the levels or changes in a predictand variable in locality (i) is forecast in terms of the levels or changes of either the predictand variable or other variables in other localities $(j, k, ...n)$. If we continue with our example of forecasting employment changes, then we find that the business cycles of sets of small areas may be related in a complex way. Some areas tend to move through a sequence of peaks and troughs somewhat ahead of most of the others. Conversely some tend to lag some months behind the others. Special interest has focussed on the "early" areas, in the hope that they might provide some early warning of recessions ahead. (Bassett and Haggett, in Chisholm, Frey, and Haggett, 1971, pp. 389-413). Studies which have attempted to isolate such leading areas have however produced conflicting results. Investigations of the cyclical pattern of a number of American cities with different industrial structures over a 26 year period (1919-45) found that though wide differences in turning points were exhibited by the various cities for relatively minor cycles these differences were inconsistent in terms of leads or lags from one minor cycle to another. During major cycles, turning points tend to be coincident. However, there was some weak evidence that cities concentrating in lead industries (e.g. Cleveland and Detroit with their steel and automobile industries) did lead other cities for most of the period investigated. Thus work in the 1950's suggested that differences in turning points among regions within the United States had not been generally demonstrated. One exception provided by a comparison between the rest of the United States and California which showed that economic activity recovered more rapidly and more fully in the latter area, suggesting that high-growth areas may show significant leads in the recovery phase of a cycle, even though coincident in the other phases.

The pessimism of the early studies has been somewhat countered by more recent studies. A study of employment in 33 mid-Western cities (ranging from W. Pennsylvania to Wisconsin) from May 1960 to September 1964 showed that a group of cities around Pittsburgh regularly lead the Detroit and Indianapolis area by 3 to 5 months. (King, Casetti and Jeffrey 1969). Similarly, regional unemployment data for ten British regions over the period 1952-63 found the Midland region leading most other regions by three months, Scotland and the North by six months (Brechling 1967). Studies of the spatial pattern of leads and lags for 60 small sub-regions in South West England in relation to turning points in the economy of the regional capital of Bristol showed peripheral rural area tending to show consistent lags while specialist industrial centres (particularly those specializing in the motor industry) tended to show lead effects (Bassett and Haggett, in Chisholm, et al 1971, 389-413).

These results for local and regional employment reflect of course the spatial structure of industry and the consequent sensitivity of each area to national and international swings in the pattern of economic activity. The concept of leading and lagging areas is however a very much wider one and geographers studying the diffusion and spread of adoptions have found that certain areas are regularly well ahead in adoption, others are resistant and laggardly. If we combine the idea of hierarchic relation discussed in the preceding section and the least-lag concept we have a combined spatial picture of the variations in one small area being linked with the variation in both the areas round about it and the larger areas of which it forms a part.

Spatial interpolation and extrapolation

In our discussion so far we have assumed that we have a set of past values at locality i which allows direct comparison with values at other localities. We can readily imagine cases where such records do not exist there and where therefore such comparisons are not possible. If we are constructing a new dam at i and wish to forecast the likely pattern of precipitation values, we are very unlikely to have any existing records for the site. How can we proceed in the absence of such records?

One approach to the problem would be by projecting future values of the predictand variable at the other locations and using these values to interpolate or extrapolate values for i. Interpolation problems are a familiar problem in contour mapping and a wide range of literature on smoothing, filtering and trend-surface mapping exists. If however the site lies outside the convex polygon formed by the outer set of points we are faced with the more difficult extrapolation problem. Krumbein (1966) has carried out an instructive test of both of the main trend-surface models – the polynomial model and the Fourier model – and finds both have severe limitations as an extrapolating tool. The general commonsense rule is that the error increases both with distance from the boundary of the reference set and inversely with the size of that set.

Evaluating spatial forecasts

It is one thing to make forecasts, quite another to make reliable forecasts. Here we look at some of the issues involved in evaluating and improving forecasts.

Evaluating short-term forecasts

Forecasts are supposedly judged on the basis of their accuracy. Thus a weather forecasting bureau may issue regular verification checks giving the percentage score of correct public forecasts: one may know for example that the daily forecasts for Chicago's weather rated around 80 per cent in 1942 improving to 86 per cent in 1963. Such general scores conceal of course important variations in the individual elements; temperature forecasting has a much higher score than precipitation forecasting since the latter has a far more broken space-time trajectory compared to the rather smooth changes in temperature levels. These inherent contrasts mean that the score of any individual forecasting method must be judged in relative rather than absolute terms.

One such relative method is to compare the accuracy of any given forecast method with "control" forecasts. These controls may be generated by drawing random values from the frequency distribution of the predictand. Alternatively the value at time $t + 1$ may be simply equated to that at time t, i.e. the forecast for to-morrow's noon temperature is that at noon to-day. A forecasting model must show a consistently better performance than such random or repetitive controls in order to be considered useful (Nagar 1962).

Evaluating long-term forecasts

Now such "wait and see" methods may be valuable for evaluating short-term forecasts or even long-term forecasts if there is a long history of testing experience to draw on. But they are patently irrelevant if we are attempting long-term forecasts in a new field. There are two major criticisms. First, the verdict is only produced *after* the results have happened. But if we are dealing with a long-term forecasting model we wish to be able to judge its likely value now, not in A.D. 2025 or whatever the relevant forecast date is reached. Second, all trend projections are based on the assumption that conditions in the forecast period are substantially the same as those in the period over which the trend model was calibrated. External changes may invalidate the particular forecast without impairing the value of the same model used in a more stable situation. This applies particularly to socio-economic forecasting where stationarity assumptions are more difficult to sustain.

For these reasons we need to develop tests that are applicable *before* and not after the forecast is made. These tests depend particularly on the internal logic of the model, that is whether or not it seems to make reasonable sense in terms of what we know about the situation. We may also use the series itself by arbitrarily separating it into two halves and using the one to project the other. The relative performance of different projection models on halved series gives us some clues to their likely performance in real forecasting conditions. These internal tests using split series are termed Janus tests (Gadd and Wold, in Wold, 1964) after the Greek doorway god with two faces, one looking backwards and one forwards. A number of biases exist in such a test and its usefulness is still in question (Granger 1971, 3).

Trend and turning-point errors

Since forecasts contain an error element it is important to reorganize the composition of the error term. Dutch econometrician, Theil (1965) has advocated the use of a simple "prediction-realization" diagram which plots the forecasted changes along one axis of the graph against actual changes on the other. Clearly an absolutely accurate series of forecasts will fall on a diagonal straight line. Errors (i.e. points *not* falling on this line) can be divided into three components: errors of overestimation, errors of underestimation, and turning-point errors. Study of the pattern of errors in relation to the Theil diagram suggests ways in which bias is occurring, and hence ways of improving the model to reduce such errors.

Theil's approach focuses attention on the apparently naive question of "exactly what do we forecast?". If we assume that the answer is given so far as the particular variable is concerned — pollution level, unemployment rate, air temperature, or whatever — the question of the *form* in which we try and forecast the data is not a trivial one. Should we forecast pollution levels or changes in pollution? Why not the logarithm of pollution or proportional change?

Granger (1971, 6) agrees that we should forecast *changes* rather than levels, since these are more difficult and give the forecasters greater degrees of freedom. A strong secular trend may be relatively easy to forecast using a black-box projective model; changes in growth may show greater instability and discourage any premature tendency to feel we understand what is evolving. After all it is not the trends which are commonsense and "surprise free" (in Kahn's scenario technology), but the unexpected turning points that provide the real challenge whether it be changing stockmarkets or snowlines. By "recognizing" turning points we refer here to the whole process, i.e. ". . . beginning with the vague early warnings forecasters usually give that the cyclical situation may be changing through the successive stages of increasing awareness until they finally confirm that a turn has definitely occurred (Fels and Hinshaw 1968, 3). Reformulating data into a form where we can evaluate our model's capability to recognize turning points is likely to be harder but far more rewarding than trend projections alone.

Length of the forecast period

As we try to project further ahead the band of probable error widens. We may use this information to throw light on the perplexing question of how far ahead we should try to throw our estimates. Let us assume that the further ahead in time a method allows us to forecast, the greater is its potential usefulness. This utility will however be limited inasfar as long-term forecasts in one sector might outrun the forecasting ability in another sector; also very long-term forecasts would raise some ethical and practical difficulties of one generation making decisions for generations very far ahead. The joint utility of a forecast is therefore a compromise between this convex utility and the concave curve of the likely error. Clearly there must be a point at which the projection time is at an optimum; beyond this optimum there is a progressive fall off in utility until a second point is reached where the errors begin to outweigh any benefits.

It is easier to outline the theoretical form of the utility curve than it is to evaluate it in practice. Where detailed studies have been carried out, the forward throw of the forecast is indicated to be very limited. For example, five-day forecasts of temperature in the United States are wrong about 30 per cent of the time, monthly forecasts 40 per cent, and seasonal forecasts nearly one half the time (World Meteorological Organization 1962). Bassett and Haggett in Chisholm, et al, (1971) are sceptical about using their lead-lag models for unemployment forecasting more than about three months ahead.

Conclusion

As Bertrand de Jouvenel points out in Art of Conjecture (1967, 5) knowledge of the future is a contradiction in terms. Only the past and present are strictly knowable: the future is still fluid, capable of being shaped in different ways. The best we can hope for is informed guesses, and only insofar as they help to inform are sophisticated forecasting models relevant or useful. In their tentative steps along the road to building more useful spatial, ecological and regional models, geographers are moving away from the simpler ground of building descriptive models. Models are to use, not to believe, and the 1970s may see the scar-marked model builder of the last decade moving forward into a more testing and rigorous area where the stakes are higher and the penalties more severe. Certainly the period of "sliding on the easy ice", as Christopher Fry puts it, is now over.

References

BERRY, B.J.L. (1971) The geography of the United States in the year 2000; *Transactions of the Institute of British Geographers* 51, 21-54.

BRECHLING, F. (1967) Trends and cycles in British regional unemployment; *Oxford Economic Papers, New Series* 19, 1-21.

CHISHOLM, M.D.I., FREY, A.E. and HAGGETT, P., (Eds.) (1971) *Regional Forecasting*; (Butterworths, London).

COLENUTT, R.J. (1972) Building models of urban growth and spatial structure; *Progress in Geography* 2, 109-152.

EHRLICH, P.R. and EHRLICH, A.H. (1970) *Population, Resources, Environment Issues in Human Ecology*; (Freeman, San Francisco).

FELS, R. and HINSHAW, C.E. (1968) *Forecasting and Recognizing Business Cycle Turning Points*; (NBER, New York).

FRASER DARLING, F. and MILTON, J.P., (Eds.) (1966). *Future Environments of North America: Transformation of a continent*; (New York).

GRANGER, C.W.J. (1964) *Spectral Analysis of Economic Time Series*; (Princeton University Press, Princeton).

GRANGER, C.W.J. (1971) Evaluation of forecasts; *Nottingham Forecasting Project Note* 1.

GREGG, J.V., HOSSELL, C.H. and RICHARDSON, J.T. (1964) *Mathematical Trend Curves*; (Oliver & Boyd, Edinburgh).

HAGGETT, P. (1972) *Geography: A modern synthesis*; (Harper, New York).

HALL, P. (1969) *London 2000*; (Faber, London).

HEILBRONER, R.L. (1959) *The Future as History*; (New York).

JANTSCH, E. (1967) *Technological Forecasting in Perspective*; (D.E.C.D., Paris).

JOUVENEL, B. de (1967) *The Art of Conjecture*; (Weidenfeld and Nicholson, London).

KAHN, H. and WIENER, A.J. (1967) *The Year 2000*; (New York).

KING, L.J., CASETTI, E. and JEFFREY, D. (1969) Economic impulses in a regional system of cities; *Regional Studies* 3, 213-18.

KRUMBEIN, W.C. (1966) A comparison of polynomial and Fourier models in map analysis; *Northwestern University, Department of Geology, ONR Contract 1228 (36), Technical Report*, 2.

LOWRY, I.S. (1965) A short course in model design; *Journal of the American Institute of Planners* 30, 158-66.

NAGER, A.L. (1962) Statistical testing of the quality of forecasts; *Statistica Neerlandica* 16, 237-47.

PATERSON, J.H. (1970) *North America*; (Oxford University Press).

PLUMB, J.H. (1970) *The Death of the Past*; (London).

PRINCE, H.C. (1971) Real, imagined and abstract worlds of the past; *Progress in Geography* 3, 1-86.

RYAN, B. (1971) Geography and futurology; *Australian Geographer* 11, 510-21.

THEIL, H. (1961) *Economic Forecasts and Policy*; (North Holland Publishing Co., Amsterdam).

WILLATTS, E.C. (1971) Planning and geography in the last three decades; *Geographical Journal* 137, 311-30.

WOLD, H., (Ed.) (1964) *Econometric Model Building*; (Amsterdam).

WOLMAN, M.G. and MILLER, J.P. (1960) Magnitude and frequency of forces in geomorphic processes; *Journal of Geology* 68, 54-74.

WORLD METEOROLOGICAL ORGANIZATION (1962) The present status of long-range forecasting in the world; *World Meteorological Organization Technical Note*, 48.

11 · Future geographies

W. L. GARRISON

Introduction: the issue

Geography is a subject of great breadth and depth involving an understanding of systems and of the relationships among their elements. It is a way of understanding man in a matrix of human and physical relationships and interrelationships. These features notwithstanding, there is a need for geographers to work at an even general and more meaningful level that encompasses the feed-forwards and feedbacks of relationships in the understanding of past and present geography, and in the shaping of future geographies. It is this need and its implications for improving the human condition that is the subject of this essay.

Such a discussion of the levels and purposes of geographical knowledge, and an attempt to predict the characteristics of a fuller geographical knowledge, is necessarily general because that fuller knowledge does not yet exist and must be imagined. It is also general because, while geographers have already achieved a high level of understanding, that achievement is patchy and fragile. Much more needs to be done even to fulfill our current aspirations and to clothe with flesh the bare bones of our existing understanding. It may be asked whether the title of this essay is meant to indicate the future philosophical, methodological, and technical content of geography or whether it refers to future real geographical systems. With the exception of "the polished present", which will be discussed immediately below, we judge the term to have both meanings and that the two are identical; for the future philosophical, methodological and technical content of geography will, together with other knowledge, determine the real geographical systems of the future.

The exception, which we have termed "the polished present" is an important one. It is a predicted future world in which, in the absence of improved understanding of geographical relationships, we merely extend the present path of development into the future. This concept applies to short-term planning because extant institutions, technologies and resource allocation programs conduct environmental improvements simply by improving and refurbishing present conditions. It hardly applies in the long run, for the system which we are presently polishing contains within it such self-destructive features as population pressures and unworkable conflict resolution mechanisms together with numerous other features that are system-destructive.

For an example of polishing the present condition, consider programs for air pollution control in the United States. Automobiles and trucks are being fitted to dampen their polluting characteristics and sources such as power stations and steel mills are being fitted with devices largely to capture their pollutants. 1975 and 1980 may show marked improvements in air quality. *May*, because there are no workable conflict-resolution mechanisms to adjudicate between clean air and economic growth. But beyond those dates, the situation begins to worsen again because there has been no basic change in the intensive use of energy, and the quantities involved will overwhelm us. We may polish the environment at great expense without having solved the long-run energy problem, because we have not addressed ourselves to the basic issue of future settlement form. There are numerous similar examples of past commodity price agreements which yielded short-run improvements and long-run economic deterioration. Later, we will refer to this style of polishing as the 'technological fix'.

The alternative to extending the geographical present into a polished short run is to develop broader knowledge leading to potential *new futures*. It is obviously easier to say what these futures should not be and to suggest some philosophical and methodological priorities that might aid in determining what that broader knowledge might be and how the new futures might be built, than to predict them in detail.

A geographical system has a past and a present, but it may have more than one future. Let S_t represent its state at a point in time, defined as a function of the relationships within the system at that time (R_t) and in previous times ($R_{t-1}, \ldots$ etc):

$$S_t = f(R_t, R_{t-1}, \ldots)$$

Our need is for a geographical understanding which is deep enough so that S may be shifted from its present trajectory, the polished future, to one of a set of alternative trajectories. For this we must have knowledge:

1. Of the possible set of alternative trajectories.
2. So that we may choose among alternatives within that set.
3. So that the processes of synthesis and action may be applied for the realization of one or more of those alternatives.

The relationships (R) are imagined to reflect a mixture of physical laws, the nature of technology, and economic, political, and social institutions. Physical laws are unchangeable, although they may be refined and supplemented, however, the others are alterable. In seeking changed relationships with their consequently new trajectories, priority must be given to identifying those feasible trajectories that may be achieved by augmenting and/or by altering existing relationships. Certain relationships are more difficult to change than others, and improvements in our knowledge are needed for us to understand how these changed relationships might lead to changed system trajectories into the future. Equally important is the problem of choice between possible trajectories. The achievement of new trajectories would be costly if they destroyed relationships which are valued and strengthened those which are not. Desirable new trajectories must be clearly superior to the goals of the polished future. The alternative futures associated with these trajectories must be closely related to the aspirations of individuals and groups, and must evolve in consort with the evolution of those aspirations.

Are new futures possible?

New futures represent departures from the future we are now creating, and they may be achieved by altering certain system relationships. Some might question whether the existing mixes of technology and economic, social, and other institutions that represent our material culture are alterable. The answer to that question would seem to be affirmative, on the evidence that these relationships have been subject to constant synthesis and change.

One should not forget that the economic, social, and other institutions forming the matrix of our culture were, for the most part, created to solve certain specific kinds of problems. The United States Constitution, for example, was created to avoid the two scourges of government by the people — tyranny or anarchy. Political, social and economic institutions may also be thought of as problem-solving mechanisms. Likewise, banks, distributing firms, and manufacturing firms solve problems of the money supply and of the production and distribution of goods and services. Such institutions are constantly being changed or overlapped by the creation of new institutions for problem solving. Indeed, one way to think of culture or the social matrix is as a set of overlapping institutions created to solve problems by establishing the guidelines within which decision-making can take place.

The question of alternative futures poses a problem of choosing between the polished present and one or more possible new futures. The manner in which institutions are created and altered is clearly central to the question of creating new futures. The present state of affairs may be largely summed up by the term "the technological fix". We currently use the technology of institution-creating largely to counteract problems as they emerge. So we support cotton prices to keep farm incomes up, we establish sufficiently high air quality standards to keep people from grumbling, we place mufflers on automobiles to reduce noise, and we create a bureau to improve highway safety. We largely use this "technological fix" to combat what is wrong without inquiring how to change the entire system so as to produce something new and better. Thus, the subsidy to cotton farmers goes on and the system continues to require subsidies. It is easy to pass prohibitory laws; to create new courses and methods of social action is more difficult, although possible.

Thinking about the future

The challenge facing us is to understand the future in the context of alternatives. There are many conventional ways of viewing the future. Some planners, for example, project their preferences into utopian worlds for all; others proceed systematically through assessments, projections and evaluation of the impact of technology. The modern utopian syndrome appears most frequently in the works of planners: in sketch plans, artists' drawings, and other descriptions of future places, for example, "East Cupcake in the Year 2000". The term "ego-tripping" may be used (and not deprecatingly) for this kind of projection, rather than a knowledge-based projection(s). It indicates that these futures closely reflect projections of the eye and the values of the beholder, such that East Cupcake in 2000 has no junk-yards and requires dense or closely packed populations depending upon the values of the planner. Designs for future transportation systems do not include places to store vehicles, and trees are always in leaf and at the proper size to complement buildings and open vistas. A feature of this approach to planning is the failure to answer the question "Can you get there from here?", for, while it may be possible to get from the present state to the future represented by these projections, how it is to be achieved is not part of the planning. Another shortcoming is the failure to answer the question "Who wants it?". Other than those of the planner, there is no understanding of whose aspirations are met by the plan. Ego-tripping is a way of life for we all tend to make gross assumptions about the values of others and to foist our values upon them. What does seem clear is that it is not a substitute for understanding the real nature of the system and, thus, its range of possible future developments.

The science fiction interpreters of technology have perhaps been the most successful in projecting the future. Their basic assumption is that technology marches in relentless scenarios, the most usual of which is a progression of the foot-in-the-door, gradual adoption and system-change stages. Thus, a technology is developed for some specific purpose, based upon existing knowledge, and in time is improved and applied in a similar way to what may be very dissimilar activities. Finally, the systems within which that technology has become embedded adapt themselves to it. In the past it has taken perhaps some fifty to seventy years to move from the first to the last act in the drama, but this sequence of change is speeding up dramatically. For example, the computer was developed to perform a rather specific job, was improved by orders of magnitude and found functions in numerous and different kinds of systems. Now systems are beginning to adapt themselves to the computer, and this has produced all manner of unforeseen difficulties including loss of individual privacy. It is only through a real understanding of how technology works that one can develop the ability to forecast the full implications of the seemingly small initial steps by which knowledge is joined in potentially-useful ways. What is lacking in existing predictions of the future is an understanding of technology creation and adoption sufficient to enable one to assemble knowledge in a context which will blend with extant and emerging social aspirations to produce matrices of alternative system trajectories. Even the technological understanding of the expert is lacking in such evaluation and future choice in a broad sense.

There are several new or recently strengthened activities that attempt rationalizations of decisions in the light of the future involving technology assessment, technology forecasting, planning, programming, budgeting and other such decision-making devices. A feature of this work is that it deals with narrow criteria for decision-making. Questions are asked with respect to the optimal decision given the present set of relationships and their path, and a yes or no decision is made from some aggregated criteria that characterizes the present system. While at the technical level this work is often guilty of decision-making based upon static relationships (e.g. making a decision about technology on criteria relevant to its "foot-in-the-door" stage), its deeper failing is that it assumes that present relationships and aggregated measures of values associated with the present system are the relevant criteria for decision-making. It is thus not set within the context of an understanding of possible alternative futures and aspirations. Because it deals only with the present and not with what might change, the history of systematic evaluation is not a good one. Such assessments, for example, have demonstrated that bombs would never be carried by rockets, that airplanes would never fly faster than 350 miles an hour, that television would have no effect upon human behavior, and that computers will never play much of a role in instruction. Until recently, as a result of similar reasoning, it was widely held that an increase in gross national product helped everyone! Finally, there are systematic projectors or prophets of all sorts from Cassandra to the Utopian, from Alexis de Tocqueville, Lenin and Herman Kahn to those who make population projections and produce urban designs to fit them. Their merits depend on the extent to which they reflect the relationships within the system and make allowance for the occurrence of the unpredictable. Such projections, however, are merely polished futures and extensions of the present paths of development. They are not linked to dynamic aspirations which might lead to new futures.

It is easy and unfair to be critical, because we have criticized ways of thinking about the future from a vantage that existing planners have not attempted to use. It is also unfair because lumping all thinking about the future into simplified categories glosses over significant exceptions. These exceptions are well reviewed by Erich Jantsch (1971) who comments upon works that he terms the "New Testament". Considering the "Old Testament" as imaginative forecasting Jantsch (1971, 68) characterizes this New Testament as "...the proper use of man's imagination in the creative (meaning action-oriented) process of restructuring human relations through changes in values, and norms. . . It bears on the precarious issues of freedom, morals, and what may be called human salvation through reason."

Creating future geographies

The issue of future geographies, therefore, involves a broadened philosophical, methodological and technical understanding, and is concerned with whether the world has but a single future, a projection of the past on its present trajectory, or whether it possesses alternative futures. We must now turn to the methodological and technical content of such future geographies.

Our understanding of relationships is based partly upon what we may imagine and partly upon what we may perceive and measure. Imagination is coupled to perception, and perception to imagination. We have spoken of alternative futures, alternative trajectories and alternative sets of relations, and of the ways in which the aspirations of individuals and groups would be met in these alternative contexts. It is apparent that it is one thing to make a general statement in favour of this way of thinking and quite another to develop the complementing methodologies and technical information necessary to create these alternatives. A basic problem, of course, is that we lack a range of environments within which we may study alternative relationships, measure alternative performances, and monitor the dynamics of alternative changes. We lack settings for creative change. In short, we do not know how different futures might perform, who wants them, or how the system dynamics might be modified either to achieve them or to improve them once we achieve them.

How to handle this deficiency is straightforward in theory, although difficult in practice. One must proceed in a systematic fashion to change the dynamics of the present system so that alternative trajectories are created. It is further to nurture the futures created by those trajectories, to observe them, to imagine how they work, and to monitor the associated changes in human aspirations and in their achievements. It is for society and creative individuals, in turn, to use these understandings in the development of its futures. In short, an experimental attitude is called for in recognizing that the present and evolving world is but one observation and that decisions about futures require multiple observations. In addition to multiple observations, experiments are commonly executed as independent trials, the trials come to an end at some fixed point, and there are calculi for the evaluation of the experimental results. The use of experimentation in predicting geographical futures requires that the methodology of experimentation be revamped and extended because the experiments that are practicable cannot possibly meet many of the conditions that the term "experimentation" commonly predicates.

Experiments may be thought of as falling into two categories. One is the result of the creative thinking through of the elements that it is practicable to put together in given geographical environments, together with the monitoring and decision calculi that will be applied to those environments. The emphasis placed upon existing and changing aspirations in the foregoing discussion is suggestive of some of the measures to be monitored. The emphasis upon relationships and upon feedbacks and mechanisms for accomplishing change are also suggestive of the elements to be manipulated and monitored. In addition to the management of technology which comes immediately to mind, new government forms, changing patterns of ownership of resources, telecommunications, and institutional changes would enter into such experiments.

An incomplete discussion of telecommunications may provide an example. By way of background, one should observe that there is nothing new about telecommunications. The telephone and the telegraph have certainly been around for a long time, and television is a familiar phenomenon in much of the Western World. What is new is the linking of computers to communications systems, the replacement of air transmission of television signals with cables with enormously increased capacity, and the imminent solution of network-switching capacity problems. Switching has been the expensive block to wider use of telecommunications. Now think about the systems in which telecommunications are embedded. For one thing these systems may be thought of as being in meta-equilibria, dependent upon technology and the social response to that technology. The system of shopping from the household for commodities is an example. There are stores of various types spaced around in certain geographical patterns based upon viable service area sizes and travel patterns. Many of these patterns have only recently evolved from earlier meta-equilibria based upon strictly neighborhood relationships. Think of shopping by telecommunications, how might this move the goods distribution systems into new meta-equilibria?

This is an example of substitution of one thing for another (shopping by telecommunication rather than by walking and looking) and the entire range of substitutions is another point to be kept in mind. A third point to be kept in mind is that of new systems characteristics and new products of the system that change might make possible. New kinds of cultural activities and self-actualization paths for individuals might be examples.

This kind of analysis stresses systematic ways of delving into how the system operates and the posing of questions in the form "what if this or that happens". Equilibrium changes, substitution, and new activities are to be sought. The example was too skimpy, but it should be sufficient to emphasize needs for thinking creatively, for carefully structuring the pattern of thinking, and for carefully thinking through the patterns of impacts of changes.

A problem is that this approach is limited to either imaginary situations or to actual but rare experiments. This kind of creative thinking could be butressed by an experimental capability to deal with unforeseen situations. Some events such as post office strikes, newspaper shutdowns, power shortages, floods, earthquakes and so forth, occasionally provide major departures from present ways of doing things. These give a glimpse of radically different environments, if only for a short period of time. Such an experimental approach as I advocate should result in a watch for such events, together with a capability to monitor and evaluate them. This same capability might be used to evaluate our storehouse of comparative geographical studies, but, because those studies were not originally carried out as part of a search for clues to creative change, it may be difficult to draw inferences from them.

The recent Alaskan earthquake sharply affected for a considerable period of time conventional ways of doing business and living in the city of Anchorage. But so far as we know no one bothered to look at how people reacted to these changes and how the rebuilding of the city might have reflected new aspirations. It is not necessary to look to disaster to find changed situations. The massive electrification schemes for railroads in the UK occasioned periods of stress and adjustment about which we know little. Also, the changing of several American cities to wider, metropolitan forms of government induced periods of change and adjustment from which we have learned little. The general admonition is simply to look and learn from those instances where systems are undergoing adjustment. And there is example after example of opportunities lost because we were not motivated to try to learn and because data collection systems and systems models were not at hand to serve as resources for investigation. A holiday weekend or a fire downtown provide grist for the mill for the observer who is thinking about system change.

The recommendation that we should both experiment with reality and monitor unusual situations, together with the caution that conventional comparative geographical studies may not be useful in this respect, are not suggestions which were attractive to most contemporary geographers. Experimentation and monitoring require scales of effort, resources, and organizations that are not readily available – nor likely to be. But there are, in fact, real opportunities for exciting and rewarding work. The first thing to do is broaden one's image from what is and what will be, to what is and what might be. At the same time, imaginative ways of thinking about what might be are in order. Thinking is needed that is imaginative in the sense that it is coupled with relations and aspirations, and that is imaginative and creative in that it seeks feasible and improved futures. Thinking in experimental formats is also helpful even if formal experiments may not be mounted. Students and teachers in a learning environment should continually balance their "what if?" questions with "what if not?". They should imagine new relationships and should think through constructive, action-oriented ways to achieve and to evaluate what is new. While complete experiments are not at hand, imaginative thinking about relations and change provides potentials for creativity.

Closure

This simple essay and its central theme – the extension of present geographic understandings to a more general level – was preceded by many personal efforts that identified facets of the issue. A decade ago, I thought that simulation was the touch-stone to alternatives (Garrison 1963). It later occurred to me that one might postulate a future and find ways to achieve it (Garrison 1966), and in this respect the availability of better information (Garrison 1965a and 1965b; Garrison et. al. 1966) and the rationalization of technology (Garrison and Schofer 1969; Garrison 1970a) have been central. Recently, the role of experimentation (Garrison and Schofer 1969; Garrison 1968) has seemed more and more important, as has institution formation (Garrison 1970b and forthcoming) and the developments of clients for change (Garrison 1969). Recently, I have traced how applications of new methods in new contexts have led to concern with the future in geographical work although we have not recognized this new concern (Garrison 1971). But all these are methodological and technical fragments of a larger challenge. They are useful, but the most useful idea is the simplest one – we must extend what we now think of as our knowledge so that it may deal with alternatives.

While our current concern with systems and relations represents a high level of understanding, it is not nearly high enough. It must extend to the alternative scenarios that these and new relations might follow, and to the problems of choice among these alternative futures. As geographers, we must recognize this concern as our central philosophical problem, and we must implement methodologies and techniques which implement this concern. Engaging in a variety of new experiments would be one method of doing this. It was no accident that the word *creative* appeared in the latter portion of the essay. Society needs creative action by creative individuals. What better stage could there be for the education of creative people than the context of future geographies and the challenge to create improved geographies?

References

GARRISON, W.L. (1963) Toward simulation models of urban growth and development; *Lund Studies in Geography, Series B, No. 24,* 91-108.

GARRISON, W.L. (1965a) Demands for small-area data; *Proceedings of the Urban and Regional Information Systems Association,* 36-40.

GARRISON, W.L. (1965b) Urban transportation models in 1975; *Journal of American Institute of Planners* 31, 156-8.

GARRISON, W.L. (1966) Evaluation of consequences of modifications to weather and climate; In "Human Dimensions of Weather Modification," *University of Chicago, Department of Geography, Research Paper No. 105,* 77-90.

GARRISON, W.L., et al. (1966) Data systems requirements for geographic research; *American Astronautical Society, Science and Technology Series* 4, 139-51.

GARRISON, W.L. (1968) Identifying alternative urban environments; In Science, Engineering, and the City, *National Academy of Sciences Publication 1498,* 93-102.

GARRISON, W.L. (1969) Innovation of new transportation systems; In *Defining Transportation Requirements, ASME,* 4-9.

GARRISON, W.L. and SCHOFER, J.L. (1969) Technology for urban systems; *Proceedings ASHRAE,* 17-23.

GARRISON, W.L. (1970a) Computer technology, urban processes and problems, new futures for urban areas; *Bulletin of the World Future Society* 3, 1-15.

GARRISON, W.L. (1970b) Coupling urban growth and development models to urban policy; *Proceedings of the American Political Science Association,* n.p.

GARRISON, W.L. (1971) *Future Knowledge*; Fred K. Schaeffer Memorial Lecture, Department of Geography, University of Iowa.

GARRISON, W.L., (forthcoming), *Coupling knowledge and decisions.*

JANTSCH, E. (1971) The 'New Testament'; *Futures* 3, 68-72.

PART V

Educational

12 · The open geographic curriculum

PETER R. GOULD

> . . . others who seemed to live in a most curious condition of consciousness, as if the state they had arrived at today were final, with no possibility of change, or as if the world and the psyche were static and would remain so forever.
>
> Carl Jung,
> *Man and His Symbols*, p. 39

> The tendency is apparently involuntary and immediate to protect oneself against the shock of change by continuing in the presence of altered situations the familiar habits, however incongruous, of the past.
>
> Elting Morison,
> *Men, Machines and Modern Times*, p. 29

If you have started to read this essay, it is likely that you are both a geographer and a member of an academic department. Hopefully, you are a student, because it means I am addressing someone who is more important in the long run. With reasonably good fortune you will be around longer, and so have more time to do something about the problems I am going to discuss. Many of them are not going to be solved with any dispatch, so I hope you start thinking about them now and take your new responsibilities seriously. If you are not a geographer, I hope you will read on a bit further. When I read about problems in other fields I can often change a word here or there — substituting "geography" for "economics", "biology" or "history" — and find that the basic sound of the argument still rings true (Morison 1966; Richter 1970). Perhaps you will find it fruitful to do the same.

I want you to imagine that you have been asked to form a new department of geography at a good, existing university, or in one that is just about to be founded. You have a clean slate to write upon, and the courses of study at both the undergraduate and graduate levels are completely open. Given the rare opportunity to write without constraint, would your curricula bear much resemblance to most of the formal courses of study to be found today? With any luck, your answer will be something like, "Good grief, No!" If your answer is something else, it probably implies that you are reasonably satisfied with the state of things today — in which case, there is not much hope for the future, and I honestly advise you to stop reading and try some other essay in this volume that may prove more congenial.

Constraints on writing new curricula

Assuming your answer is resoundingly negative, it is interesting to speculate *why*. It must mean that your perception of things as they are, and things as they could be, are seriously out of kilter. Things as they are, or reality (r), and things as they could be, or an ideal state (i), are both functions of human, organizational, and uncertainty factors. In other terms:

$$r = f(h, o, u) \quad \text{and } i = g(h, o, u)$$

In reducing the disparity between reality and the ideal state, we can think of trying to minimize the function:

$$D^2 = \sum_{k=1}^{n} (r_k - i_k)^2$$

What is preventing the minimization of D^2, your perceived disparity statistic? Clearly the function is subject to human, organizational and uncertainty constraints, and if we simplify these by gross aggregation then we can write:

$$\sum_{j=1}^{\ell} h_j \leqslant H \text{ for the human constraints}$$

$$\sum_{j=1}^{m} o_j \leqslant 0 \text{ for the organizational constraints}$$

$$\sum_{j=1}^{n} u_j \leqslant U \text{ for the uncertainty about the future constraints}$$

we can then define a new function to minimize:

$$D_\lambda^2 = \sum_{k=1}^{n} (r_k - i_k)^2 - \lambda_1 \left(\sum_{j=1}^{\ell} h_j - H \right)$$

$$- \lambda_2 \left(\sum_{j=1}^{m} o_j - O \right) - \lambda_3 \left(\sum_{j=1}^{n} u_j - U \right)$$

Where λ_1, and λ_2 and λ_3 are the Langrangians. We must assume these are negative in sign to make them binding, and we interpret them in the conventional way as the marginal rates of change we can expect if the constraints should loosen — or, Heaven forbid, tighten up.

I would be the first to admit that all this business of D^2's and Langrangian multipliers is really a bit pretentious, *except* that it does help us to focus rather explicitly upon what we are trying to do as we write our new geographic catalogue, and what the constraints are as we move in the future to a better state of affairs. It also directs our attention to two other things: first, that the constraints and the Langrangians may not be the same everywhere; and, secondly, that we can never find the global minimum of our function, because some constraints will always bind no matter what the situation. Knowing this may make them chafe a little less. At the moment we cannot go much further in a formal fashion because we do not have the data. Getting the data would be a difficult, but not I think impossible task, and one that might prove extremely illuminating on a comparative, international basis. In the meantime, and because the constraints do bind, we really should examine them in a descriptive way as a background for our discussion of curriculum writing for our new department.

The human constraints

Most people like to keep things as they are to minimize uncertainty about the future and relieve them of the obligation to do anything about it. Geographers in the academic world are probably no more conservative than others — which is to say that many are very conservative indeed. Although doctoral degrees are still awarded at commencements, many appear to regard them as terminal states.

This is hardly surprising: the *average* time between the first,
Bachelor's degree and the Ph.D. is 9.3 years (Taaffe 1970), and there
are few who can retain a vestige of intellectual curiosity after so
tedious a course. Anything new in the field represents further
gruelling effort, and may also mean acknowledging that much that
was learnt before is now useless, irrelevant, or downright trivial.
Such acknowledgement takes a rare sort of courage. The ideal
teacher and scholar — particularly the former — is a person who has
accepted the notion that he will be running all his life with a cold
feeling in the pit of his stomach that he can never keep up. Yet he
tries desperately to do so knowing, like Sisyphus, that by his own
standards he will always fail. Yet, paradoxically, his constant
acknowledgement of failure results in a minimal degree of
intellectual self-confidence that forbids him to regard the new as a
threat. Moreover, his constant efforts hone his intuition and
awareness, allowing him to encourage students to explore the
geographic utility of new ways of looking at old problems. Of
course, his encouragement is seldom purely altruistic: one of the
ways he can keep up is by having his students do some of his
learning by teaching him. Quite contrary to traditional ideas, really
good departments will be characterized by students teaching the
faculty. Any graduate student who cannot get ahead of the faculty,
and turn around once in awhile to teach the poor things plodding
along behind, should be dismissed. Students who are poor teachers
of the faculty have no place in a university, for they have little to
contribute to its life.

No teacher, whether student or faculty, is ideal, and to the degree
that he does not try to learn, does not encourage the exploration of
the new, he forms part of the aggregate human constraint on writing
an open and changing curriculum. It is difficult to loosen these
human constraints that can bind with such intellectually throttling
force. A host of questions, starting with the security of tenure, are
raised, and would require another essay even to lay them out.
Perhaps all we can do is try to set a good example ourselves, and so
minimize our effect in the aggregate constraint. Inasmuch as we are
aware of the influence of our own role, and acknowledge the
possibility of our constraining effect, we are in a position to try to
minimize it. Unfortunately, there are always some who, far from

acknowledging such a possibility, regard themselves as guardians of the "right" tradition — usually one in which they have a considerable vested interest. Sometimes they gain positions of influence in archaic organizational structures that allow them to thwart developments during, and even after, their formal professional lifetimes. For our model is really an over-simplification: sometimes the human and organizational constraints interact in deadly, and deadening, combination.

The organizational constraints

The most binding organizational constraint by far is the European professional system that emerged several hundred years ago. In the meantime the world has moved on to the Twentieth Century, where the organizational needs are somewhat different. In the same way the Mahan noted that military organizations are incapable of reforming themselves, so all the evidence points to the impossibility of entrenched academics revising a system in which they themselves are embedded. This is particularly true of most European depart-ments, where the decision-making structure is extremely hier-archical.

The European situation is especially distressing in view of the strong traditions of geographic enquiry in the past, and one senses considerable frustration building up, especially on the part of some younger geographers. The traditional aims of many European departments, to convey a body of general "culture" and to prepare high school teachers, are parts of the overall problem. Given such goals, it is difficult to bring in questions of practical import, methods for tackling geographic problems, and intriguing research findings — in brief, the very things that convey the intellectual zip and excitement of spatial pattern, structure and process. Large and plodding bodies of fact do not transmit a sense of ferment and questioning to students, and it sometimes seems that the traditional regional monograph is the tombstone of European geography.

Applied geography, which can be such a relevant channel for so many students today, appears particularly difficult to handle. Even where brave and pioneering attempts have been made, students face the difficulty of the incredibly closed nature of European pro-fessional life, and in many countries it is almost impossible to break into regional planning circles. This must be very discouraging, for often the investigation of real problems appears to provide the only channel through which new ideas can seep. Certainly the rise of Polish geography during the period of post-war reconstruction must be due in large part to the strongly applied orientation of the discipline in that remarkable country. Unfortunately, the reverse seems to be true over much of Western Europe, where the growing number of positions in spatial planning are nearly always filled by economists and engineers, whose analytical training lets them take a good cut at solving problems, rather than writing tedious essays about them. It is, perhaps, worth recording the curiosity generated by one of my own students, who held until recently a prominent analytical position in Europe's largest research and development corporation. When he applied with an American Master's degree, there was genuine skepticism about his geographic credentials. In the first place, he was bilingual in FORTRAN and ALGOL, as well as French and English, and his thesis dealt with network analysis. His record showed courses in linear programming and theory, which was hardly in the geographic tradition, and, anyway, no other geographer had ever applied for such a position before. Yet his geographic credentials are roughly those of any major today who has his first, Bachelor's degree from a good American department. Given the stultifying effect of many European geography departments, what hope is there for getting a crucial, and well-trained spatial viewpoint into critical areas of planning?

Not all is discouraging, however, and there are signs of resurgence that all geographers will welcome. In a devastating critique of the current situation, an editorial for a new French journal entitled *L'Espace géographique* demonstrates the deep concern for geography held by many. Whether this is the first sign of a new era in French geography remains to be seen, but there remains a spark of hope that might one day burst into flame. In a larger Europe, a young and vital generation of French geographers has an enormous contribution to make; not the least through cooperative research with other colleagues, perhaps supported by the Europe 2000 Foundation. After all, it is only the young and creative geographers today who stand a chance of shaping the spatial organization of Europe in the year 2000. Hopefully such a resurgence will soon characterize developments throughout the continent, particularly southern Europe, where students of geography are nurtured upon curricula straight out of the Twenties — behind half a century of rapid change and paradigmic shift (Tricart 1969).

Another constraining effect of the European university tradition and organization is the enormous amount of inbreeding one observes in many departments. Far from maintaining the medieval tradition of the wandering student and scholar, many today appear totally committed to one department from their earliest university years. The result is institutional nepotism that produces academic families whose ideas mesh, and stagnate, together. Often a faculty member appointed from the outside is regarded very coolly because it means he has usurped the place of a "family" member. Similarly, few students would think of moving to another department for their second or third degrees, because they might lose their "place" at their own institutions. No reform could have a more stimulating effect upon the geographic life of Europe than university directives forbidding students to take all their degrees at the same place. Few American students can imagine the genuflections required of the European student who wishes to change.

On the more positive side, it is true that a European department *can* move with a speed and vigor difficult to match once the chair is held by a forward-looking professor. Several examples in England are well-known to all. A chair holder can shape a department relatively quickly, and move programs of study along more exciting contemporary channels with a speed unmatched by the more open and democratic departments in the United States. But such surges usually represent a catching up with the best American departments, and for sustained growth these are proving difficult to match. In virtually all the good departments, almost by definition, the chairmanship is regarded as a burdensome obligation to be shouldered after much armtwisting by less responsible colleagues. No one in his right senses would assume such a position as a lifetime obligation. Most decisions with any bearing upon curricula and training are made in open faculty meetings, and while there may be blood all over the walls on occasion, the sessions do provide an open forum for new ideas. It is not a perfect system, but I have little doubt that the values of most of the American Lagrangians in our equation are considerably smaller than those of Europe.

The more open American system, with its large number of

universities of all gradations, does suffer from problems of instability, however, for the mobility of faculty tends to be much greater than in Europe. While the stability of stagnation or smugness is to be avoided on the one hand, the game of academic musical chairs can be equally devastating when carried to extremes. Good programs are built patiently by reasonably stable groups of men and women, and while the play may be more exciting for the actors if they constantly rush in and out from the wings, the student audience may become bewildered and lose the plot. It is always an unhappy business when a student loses his graduate advisor, with whom he has built up a relationship of mutual respect, intellectual sympathy and concern. Perhaps the answer is to build as much as possible with young scholars, from the bottom up, to maintain a stability that at its best is a delicate balance.

Constraints from uncertainty about the future

Whether we try to revise existing curricula, or write new catalogues totally free from human and organizational constraints, we must always face the problem of uncertainty about the future. Charged with the responsibility of devising curricula at the undergraduate and graduate levels, what can we teach today that will help tomorrow? It is an almost impossible question to answer, but perhaps we can think about some possible criteria for selection that will help us choose and decide as pendulums of fashion swing more widely around us.

One thing, I think, is perfectly clear: the intellectually stultifying effect of many geographic courses is caused by their large and usually unstructured *factual* content. Great choking gobbets of facts are purveyed in traditional courses, and it is hardly surprising that they stick in the throats of students — particularly the more intelligent ones. Learning large quantities of facts, and being required to regurgitate them by multiple-choice examinations (U.S.A.), or by waffling ignorance-cloaking essays (Europe), is not going to help students tomorrow. Today's and yesterday's facts are the last priority, at the bottom of the pyramid, to be dipped into once in awhile to illustrate . . . well, to illustrate what? Surely, a body of theory and conceptual insight that *may* help a person reach, after suitably demanding mental sweat and discipline, a higher state of awareness and insight about human existence in the two dimensions of geographic space. No one is pretending that these concepts are easy, or in all cases well-formulated, but it is a body of theory that should form the core of curricula at *all* levels, because it stands a better chance of being of greater intellectual worth and practical utility in the future. Thus, it helps to loosen the binding uncertainty constraint.

But the constraint of the future can be loosened still further by taking students even higher up the pyramid where they can learn analytical approaches to solving geographic problems themselves, obtain the presently available tools of analysis, and acquire a body of very general, deep and powerful concepts that cut across all the sciences. As certain as things can be about the future, it is certain that problems of maximizing and minimizing (hopefully subject to constraints not considered today: Harvey 1971), of stability and equilibrium, of dynamic paths and spatial forecasting (Chisholm, Frey and Haggett 1971), and of handling large and complex systems will be around in any future we can see. A curriculum that opens up these very general problems to students, and displays them in that spatial context for which geographers possess such intellectual passion, is surely a curriculum that has itself minimized the Lagrangian of the future constraint.

Pendulum swings are difficult to cope with, and if enough people hang on the momentum can move a whole discipline into new paths (Kuhn 1962). Sometimes such changes are healthy, but it is often difficult when one is being battered by a large pendulum to know whether a paradigmic shift is coming around the corner, or whether just another bandwagon is rolling by. A mistake in judgment can be unnerving. Despite similar shifts in most of the physical, biological, behavioral and social sciences, a number of geographers dismissed the rise of quantitative methodology in the late fifties as a fad, when it really signalled a major shift in the field. Some responded by moving into evermore esoteric 'qualitative' niches, pursued by younger colleagues and students who measured the unmeasurable, and related the unrelatable. Others responded with a we-were-doing-it-all-alongism, which turned out to be an old-hatism really too shabby to be worn in polite company. Even when tricked out in hastily-sought classical quotations, to give them an air of scholarly verisimilitude they would otherwise have lacked, such responses have been less than helpful (Crist 1969).

At the present time the pendulum is swinging towards a behavioral geography, characterized very strongly by a concern for the micro-spatial behavior of the individual, his cognitive mechanism and psyche. Some would argue, and very persuasively (Hägerstrand 1970), that only by going down to the basic components of human geography can we ever hope to build sound geographic theory. Yet I am not wholly persuaded that this is the course all should follow by any means. If the history of all the *micro*-fields should tell us anything, it is that obtaining deep insights into the individual components does not necessarily help us understand the way in which larger systems and aggregates hang together. The history of biology and its micro-subfield over the past thirty years is worth examining closely in this connection, particularly if we assume that one day, perhaps beyond our lifetimes, human geography will be absorbed as part of a larger, deeper and more powerful biology (Hägerstrand 1969). Rather I would agree with Beer that we desperately need insights all the way up and down the cones of resolution (Beer 1968; Curry 1971), a nested hierarchy of models that is articulated as carefully as we can make it. After all, it is equally easy to argue for the necessity of a macro-approach. A new social physics, with rich sources of analogy in statistical and quantum mechanics, appears mandatory whenever we encounter the literally unimaginable and unthinkable numbers of possible configurations a small human system can take (Wilson 1971). Even a sample of ten men and women moving on multiple shopping trips to four stores can generate over a trillion (2^{40}) possible configurations. In what language can we possibly write spatial theory except that of statistical mechanics and thermodynamics?

But it is not enough to lay out the apex and base of geography's cone of resolution: we must try to think through the implications of this broad spectrum of scale for our future-oriented curricula. Ideally, a well-trained undergraduate major should be able to handle much of the current research in his field. And when I say *handle*, I do not mean he should be able to read immediately with complete understanding, but he should have the intellectual training to gain at once some intuitive comprehension, and the technical equipment to deepen and broaden his understanding with disciplined effort. In brief, and no matter what his particular predilection within human geography, he needs the language of mathematics, some experience of imaginative spatial application, and much exercise in its geographic use.

Others, particularly students, would post a higher priority to release the uncertainty constraint, arguing with great force that human problems in both developed and developing countries are so urgent they should strongly shape our curricula. Few could disagree with the basic tenet of urgency: many would differ on the manner of the shaping. It is not enough to run through the corridors wringing anxious hands over man's fate, hallooing the cry of relevance and berating faculty as uninvolved and uncompassionate if they are not tramping the streets of the ghetto. Equally, it is not enough to run courses on the geography of poverty which merely describe one woeful condition after another, and end up serving as a cathartic, rather than a disciplined intellectual experience. But surely it *is* responsible to bring relevant materials and cases illustrating man's inhumanity into existing and future courses, to show that many of these problems are not going to be solved by hand-wringing and rhetoric, but by tough, sustained and disciplined analysis — in brief, the very thing that any university curriculum should provide. The problem with students shaping a curriculum is that they are esentially shortsighted and basically conservative, especially when attempts are being made to raise the very intellectual standards by which they may be judged. The trouble with much of the descriptive material on poverty and social injustice is that it is easy to read and understand; it is not intellectually demanding or stretching, and should form background material rather than substitute for analytical substance.

Finally, there is the great difficulty of judging what will be relevant in the future, and our experience of past attempts should make us extremely cautious about dismissing any area of inquiry as irrelevant. A few years ago, academics who studied ants, wasps and other social insects hardly appeared relevant to the great and pressing needs of society. Today we realize that these fauna represent an enormous biomass in our ecological system, turning over, digesting and reducing more material than any other group.

Wipe out mankind and the ecology of our planet would jog along
with only the brief tremor of a large, and rapidly decaying protein
input. Destroy the social insects, and the ecology of our planet
would be radically altered. But where are the demonstrators for
more funds to help us increase our understanding of this vital link in
the system in which all human society is ultimately rooted? And in
geography, who would have labelled as socially relevant a man who
patiently studied for years the formations of underground ice in
Canada's Arctic? Yet today his "merely academic" knowledge is
fundamental to the use of new energy sources and the extension of
man's livable environment into new areas of our planetary home
(Mackay 1971). Not all the relevant work is in the ghettos of
America's cities: sometimes a person sitting in an armchair with
pencil and paper can save more human lives than all the hand-
wringers together (Dantzig Orden and Wolfe 1954; Shen Lin 1965,
1970; Törnqvist 1963).

Perhaps all we can do to loosen the uncertainty constraint is to
think through the implications of our own work and teaching for the
future. In the same way we can distinguish between antiquarian
curiosity and demanding archeological study, between recreational
hobby and disciplined interest, so perhaps we can make discrimin-
ating judgements about our own curricula, cutting out with care the
less demanding branches to let in more light and air for the stronger
and healthier shoots. Otherwise we must face an agonizingly slow
process of revision by accretion; a Parkinsonian process that results
in bulging course offerings, most of which would have been phased
out decades ago but for the emotional involvement of those who
taught them over perhaps too many years.

Curricula jottings on a clean slate

Like Winston Churchill facing a fresh and gleaming canvas with
fully-charged brush in hand, I find the freedom to write a curriculum
on a clean slate a rather unnerving and trepidacious experience.
These, then, are jottings set down to delineate some broad outlines,
rather than heavily engraved strokes.

Starting points are important, and I want to consider what is,
perhaps, the most vital part of the curriculum — the first university

geography course. If this course is challenging and demanding, rather than simply time-consuming, it can set the tone and force the upgrading of all subsequent parts of the curriculum. If it is shallow, it can pull down much towards it.

The beginning course

Under the American system, and increasingly in Europe, the beginning university course serves two main purposes: as a foundation for geography majors who will build a subsequent structure of greater sophistication and deeper insight, and as a service course for students in other disciplines. Often these purposes are viewed in conflict, and new, rather dilute courses are devised for the service function to pick up all the credit hours so dear to the hearts of administrators. I would like to argue, as strongly as possible, that there need be no conflict here, and that we truly serve the university community most honestly by presenting the richness of geographic concept and theory at a suitably rigorous level to stretch both the potential majors and the curious, but probably terminal students from other disciplines. At a time when geography has everything going for it, when students particularly are realizing that we exist in space as well as time, there is little need to teach watered-down "general education" courses, which end up so diluted and devoid of challenge that they drive the intelligent students away forever. Even today, and despite the lengthy example of one of the finest textbooks ever written in the social sciences (Samuelson 1967), we see too many beginning courses that are factual jumbles, bereft of the simplest and, therefore, the most powerful concepts of our discipline.

The example of a good beginning course in economics, built around Samuelson's book, is instructive. Very simple, but powerful models are presented, and it is these that provide the coherence and order for a seemingly disparate set of facts about economic behavior. No economist is so half-witted as to believe that supply and demand curves are simply a couple of straight lines crossing on a graph, and none would deny that the estimation of these functions can end as a most taxing exercise in advanced econometrics. But these gross simplifications and abstractions provide a powerful framework for discussing more difficult ideas, and possess the great merit of all simple models — that they can easily be made more and more complicated to stretch and challenge the student. So in geography, our conceptually rich, but simplified models should be the starting point.

Nor is anything but elementary mathematics required at this point. Every student brings to the university a knowledge of arithmetic, basic algebra and geometry. Many today bring a knowledge of elementary matrix operations and the calculus. But at the beginning level, the calculus is not required, and much can be done with imaginative graphics. After all, eight-year olds in England, working with one of the most imaginative books ever written at that level (Cole 1968), are tackling problems of spatial networks, minimum travel locations, and spatial correlation measured with two-by-two contingency tables. Is it possible that we who teach and write sometimes underestimate the intellectual capabilities of our beginning students?

It is also very important at this level to give students a deep sense of the unity of all human geography, its roots in past developments, and how it fits into the broadest context of man's enquiries (Abler, Adams and Gould 1971). There are few sadder tasks than riffling through a university catalogue to find the beginning level split up into a series of courses labelled with such archaic titles as economic-, cultural-, political- and urban geography. Presumably these beginning courses were incorporated during a period of innovation in the Thirties, for they seem to reflect the focus upon lots of little esoteric facts rather than upon a body of theory and concept. What sense does it make, for example, to discuss tariffs in economic geography, administrative boundaries in political geography, linguistic constraints in cultural geography, housing constraints in urban geography, and so, wearily, on? The important point is that there are *barriers* to all sorts of human flows across geographic space, including the barrier of distance itself, suitably warped and twisted by transportation and communications technology. Show the beginning student how to handle, think about, measure and analyze barriers, and you have given him a higher-order concept that means he will never be able to think about this aspect of human spatial organization with quite the same naiveté again. At this point, all sorts of intriguing examples can be used to put some real world flesh on the conceptual bones. These can range from the bushmen with their buried ostrich eggs using the barrier of the Kalahari against pursuit by irate Bantu neighbors, through the shaping of ghetto diffusion by racial barriers, to the growth of the U.S. demand cones for strawberries into Mexico with agricultural tariff changes. There is a genuine question of intellectual *efficiency* here, a word that may make some of us in the academic world uncomfortable – until we reflect that the opposite may be sheer *in*efficiency, which is difficult to defend. There is no question that we have an area of the traditional curriculum here where much consolidation and pruning is possible, to reduce the usual overblown list while sharpening the contents.

The next step

What should be the next step in our well-structured curriculum? "Well-structured" implies that we have given some thought to the sequence of geographic instruction, as well as considering the actual contents of the parts. Perhaps the greatest problem we have failed to grapple with in the past ten or fifteen years is the divorcing of methodology from other areas of geographical inquiry. Often a course or two of something called "quantitative techniques" has been stuck in late in the undergraduate curriculum, mainly because the contents were unfamiliar rather than difficult, and because we seem to have been inexcusably sloppy when it came to thinking through the exact role of methodological instruction. The result has been that technical insight has been torn from real problems in the minds of many students, instead of being regarded as part of the basic analytical toolkit with which to tear problems apart and solve them. It is time to think through this unfortunate schism and the place of methodology in the open curriculum.

There are really two problems here. One of them involves the actual content of the quantitative courses, some of which are taught in a most unimaginative way. After the third week of drawing black balls from imaginary urns, most students are crawling up the wall – and who can blame them? Although at some point the fundamentals of probability theory must be absorbed, students should see the way in which simple approaches are geographic from the beginning. And since many of them will be afraid of numbers, something intriguing should come first. Every teacher will have his own pet approach to cut into the methodological circle and break the hold of psychological resistance, but simple regression techniques are to be commended for their arithmetic simplicity, ease of graphic demonstration and intuitive appeal. They also allow the immediate incorporation of the traditional map into the first lecture, and deal with the fundamental scientific concepts of measurement and relationship which open up a host of real-world problems to a student.

Courses of this sort will obviously reflect the individual strengths and weaknesses of the teacher, and the contents, exercises and approaches will vary considerably. But the second question — of position in the undergraduate sequence — is more difficult. Contrary to the traditional curriculum, I would argue that the methodological fundamentals should be placed as early in the curriculum as possible. Sequences imply priorities, and priorities imply that some things should come before others. I hope it will not appear too simple-minded to recall that the converse is true: some things come after others, and rest upon what has gone before. This is presumably what we mean when we say that we are building a curriculum — putting bricks on top of one another, rather than simply laying them down side by side. Having been exposed to a wide range of geographic problems in a well-integrated sequence of beginning courses, is it not logical to take up next the tools needed to solve them? In this way, elementary methods are sandwiched between beginning and more advanced work, and provide an essential bridge to junior and senior level work. The alternative is to stay with the distressingly muddled pattern we have today, in which advanced systematic work is either quite impossible, or at the least hopelessly inefficient, because the students simply do not have access to much of the current literature. Time in systematic courses is used for what is essentially remedial methodological work. One could argue that such a scheme integrates method and substance, except that the teacher will always be faced with half the students having some of the techniques from another systematic course, while half have not. So he teaches them all over again, to the boredom of a large proportion of his class.

Putting the fundamental methodological courses near the

beginning is going to make severe demands on some teachers. On the other hand, there is no reason why they cannot learn from their junior and senior students, who come to them clutching mental toolbags filled with gleaming new instruments. Conversely, it is not going to do students any harm to learn sooner, rather than later, that there are some geographic problems that cannot be solved by factor analyzing every matrix in sight. The tools will be a bit dull after some good advanced courses, but their sharpening should provide ample incentive for more demanding work at the senior and graduate level. Thus, our curriculum assumes an interleaving of methodological and systematic courses, the sequencing of which depends upon the intellectual demands we are prepared to make in the upper level courses. Many of these today are repetitive and time-consuming, rather than demanding the intellectual discipline required of all truly ground-breaking efforts. Of course, we have the implicit assumption here that if the methodological courses are the warp, the systematic courses are the woof. This is an assumption well worth examining in its own right.

Are systematic courses out of date?

It hardly seems credible now, but there was a time when courses in urban-, settlement- and transportation geography were considered quite *avant garde.* Older geographers can easily recall the fights they had in getting such outlandish titles into the catalogues, which were stuffed at the time with regional topics. Few would object today, but the irony is that just when everyone has comfortably settled down into new categories, the categories themselves have outlived their usefulness, becoming limitations upon geographic instruction and pedagogic imagination. With our slate only half filled so far, do we want to write down such old-fashioned labels once again? Remember that we have students from the first portion of our curriculum who have had an excellent introductory sequence demonstrating the unity and inter-relatedness of human geographic inquiry. They have followed this with the methodological elements of multivariate analysis, and probablistic and normative modelling. With any luck they are raring to go. At this point, are you going to make them plod through now traditional systematic courses? I think not: most of the high points are familiar to them from the beginning and methodological sequences,.and they are going to resent a simple interstitial filling. Indeed, some of the more outspoken ones may even say you are padding. What, then, is the next step?

At this point, the student should have a chance to spread his wings, and with a solid introduction and some basic tools under his belt begin to explore new themes with some relevance to his times and his own intellect. Courses should be available to him in his junior and senior years very different from the traditional systematic offerings of today. The difficulty is that we come smack up against entrenched institutional constraints, for administrators like tidy catalogues planned two years in advance, and in Europe official examination syllabi frequently enshrine the truth that students shall know. But quite simply, these constraints must be broken. For the traditional systematic offerings in our catalogue we should substitute an array of open and *constantly changing* themes whose subjects are announced at the beginning of each academic year, or even closer to the time of starting. I envisage a flexible hybrid between the present formal lecture course and graduate seminar, with the instructor acting more as an intellectual resource than a fount of wisdom. Perhaps an initial series of lectures could delimit the major theme of the course, and set out a number of interesting sub-areas, but then students should be able to explore with their own resources, backed by the university library and computer center.

Much of the important literature in the field today would still be used in these thematic courses with a strong theoretical or problem orientation, but in supporting a specific problem area it would appear much more pertinent. The problems or themes are limited only by the disciplined imagination · of the faculty or students, ranging from the geography of waste disposal and recycling (Zelinsky 1971), through the location of social services (libraries, schools, children's playgrounds, hospitals, etc.), to spatial prediction and the future geographies of specific regions or developing countries. Thematic courses can look back, and focus upon problems of geographic reconstruction; take the present and try to solve with field mapping and computer simulation a local question; or look forward to deduce the future spatial consequences of ongoing, and perhaps disastrous political and economic policies. There is no limit to the themes: all we need is a new and open curriculum with a flexibility unconstrained by archaic lead times and examination procedures.

You will have noticed that I have not mentioned three topics in most of our present curricula: physical geography, regional geography and cartography. Physical geography, as most of us knew it, either survives as a second-rate earth science or has vanished completely. I regard it as a separate field, with problems so taxing that they demand the full attention of the student if anything more than a superficial acquaintance is to be achieved. Most of physical geography is totally irrelevant to human spatial organization, except at the most obvious and naive level. One does not have to wade through the last ice age to understand the Sahara is now a desiccated and inhospitable environment to man at his present stage of technology, any more than we have to start millions of years ago to understand that the long ridges and valleys of Central Pennsylvania are barriers to human movement and communication. Moreover, in areas where specific aspects of the physical environment do impinge upon man, for example in new areas of agricultural settlement and expansion, the problems require the expertise of the soil chemist, microclimatologist and engineering technician. We do not expect the contemporary human geographer to be grounded in every conceivable field of the slightest possible relevance to him, for then we would have, quite literally, a lifetime curriculum. In fact, excising physical geography is not nearly so traumatic as many would have us believe. In Sweden, for example, there are now quite separate departments and curricula, to the strengthening of both fields and the alleviation of schizophrenia in the students. At leading British universities there is still a token first year of shared work, but after that students go their separate ways, and one has the feeling that as an older generation retires the choice will be made earlier. In the United States, physical geography has been fading slowly away for decades, though students are still forced through a certain amount of ritual. A final reason for the excision is that any relevant materials can be incorporated far more effectively and excitingly at the beginning level as examples of ecological systems. Rainfall and evapo-transpiration do vary, and we can form *a priori* probability distributions from which to draw stochastic values in a game, or linear programming simulation. One does not have to commit to photographic memory the wretched Köppen classification for that, although this scheme might be used as an early spatial example in a methodological course dealing specifically with the fundamental scientific problem of classification and grouping (D. Steiner 1965). Regional geography has also disappeared from the undergraduate curriculum for two good reasons. First, it was the area of the traditional curriculum most notorious for factual memorization and regurgitation, thwarting the development of analytical abilities by muffling them in gross memorization requirements. Most American graduate departments have had to face the unhappy problem of European students with superb degrees who were absolutely lost when it came to applying their presumably atrophied analytical talents. Even more distressing is where this factual tradition has embedded itself into the newer universities of Africa and Asia to produce students with excellent memories, but few independent and analytical abilities. In its present form, regional geography has done enough damage, and the quicker we get rid of it the better.

But the second reason appears almost paradoxical: regional geography has disappeared from the undergraduate curriculum because it is too difficult. It is within the region, or spatial laboratory, that the most complex interrelationships occur. Even if we draw our boundaries to maximize the degree of closure of the system, and only now are we beginning to do this (Brown and Holmes 1971), we are still dealing with the most complex area of spatial analysis we can conceive. We have to bring to a specific part of the earth's surface every insight and every tool at our command to show how a human system of numbing complexity moves through time. With much talent at the graduate seminar level, a group of well-trained and imaginative students might be able to handle a piece of the problem with large and specialized library resources. But for undergraduates, still in the process of learning geographic theory and methodology, and cutting their teeth on limited problems and themes, the usual regional course can only appear superficial. Pleasant enough, perhaps, to hear a good lecturer ramble around Africa or Latin America, especially if he has a good fund of personal anecdotes, but let us not confuse this with geographic training when there is already so much to do.

The third area, cartography, has not really disappeared at all. It has only shucked off its busy work of French curves and crow quill pens, and has been planted more firmly in its traditional mathematical soil from which it has long been uprooted. While a case can be made for drafting on the grounds of occupational therapy after a long period of writing, in a *university* it is very difficult to justify the hundreds of hours normally spent by undergraduates over the light table. Even today, much of the work can be done by machines at a minute fraction of the cost of hand labor, and the simple production of maps can be effectively incorporated into the sequence of computer programming required by all major departments for their undergraduates. Some of the "difficult" areas of cartography, which are usually left out because there is no one to teach them, can be handled as part of the complementary mathematical sequence that is so vital to geographic training today. Truly advanced work would appear in courses at the senior and graduate level. Mathematical competence has always been essential to any cartographic work of substance, and it is absurd trying to teach this ancient and honorable discipline to students without the basic linguistic skills. For we shall see in a minute that mathematics is a language crucial for any scientific inquiry, including the geographic.

Mathematics: cross-braiding parallel streams in the curriculum

Many people are afraid of mathematics, and with good reason. It makes difficult demands upon the intellect, and more often than not it is badly taught by people who found the elementary forms so simple and obvious that they cannot understand why anyone should have any difficulty. That a minimal degree of mathematical competence is rare among geographers should come as no surprise. Fortunately, there are encouraging signs that the next generation will be better equipped. In the BASS report on geography, for example, American chairmen overwhelmingly recommended mathematics and statistics as the most important cognate fields (Taaffe 1970), and on the mathematical side there is a growing awareness of the needs of the behavioral and social sciences (Kruskal 1970).

The reluctance to learn mathematics is frequently compounded by a feeling of irrelevance that is rooted, in turn, in the most damaging dichotomy of the western world – the split between the two cultures to which so many have alluded since Snow's original statement (Snow 1959). Even now it is difficult to persuade anyone of an enlarged view of the humanities, a view so expansive and open that it embraces all the works and curiosities of man. Few seem capable today of moving forward to the Eighteenth Century with its glorious view of natural philosophy nutured by roots deep in the soil of ancient Greece. There, at least, geometry and number were not separated from the discourse of man. Today the fear of even elementary mathematics generates deep antagonisms, and there are many who are willing to ascribe to descriptions cast in this language a dehumanizing element (Bartholomew and Bassett 1971). Intellectually defensive postures abound, and accusations fly that many geographers strive for pretension by failing to communicate. However, communication is a two-way street. Is it conceivable that a physicist, biologist, astronomer, engineer or meteorologist, faced with a difficult mathematical expression, would declare that his colleague was failing to communicate? Rather, like a good Christian, he would take off his coat and go to work, for the burden of communication would be upon him (Matthew, *circa* 50).

Implicit in such an attitude to the failure of communication is the assumption that all things may be expressed in words. The assumption is frequently false. As George Steiner has argued in one of his most perceptive essays of literary criticism, mathematics is a language, and there are areas of human inquiry and expression for which it is the natural medium (G. Steiner 1969). Since the Eighteenth Century, mathematics has become the language for expressing ever larger portions of deepening experience, and translation is often distorting to the point of dishonesty. The history of all the sciences is the same: there follows an age of patient recording another one of measuring, classifying, structuring relationships and increasing mathematical expression by simplifying modelling. In the human sciences from medicine to economics, and in the sciences of design from architecture to engineering mechanics (Simon 1969), ever-widening portions of understanding are communicated in the language that is natural and appropriate to them. Like any language, mathematics opens doors to other areas of inquiry. Not only is there a growing and important literature in geography, but a huge body of mathematical writing in adjacent fields of extreme relevance. Many have commented upon the intellectual parochialism of forty years ago, but we can see today that much of the barrier effect was due to sheer linguistic difficulties.

Am I too far off the mark if I point out the problem of motivation at the heart of our mathematical difficulties? If people can see *why* a set of basic skills should be acquired, they will have the incentive to start the hard and slogging work required of them, and the strength to keep them going over the rough places. What we need to explore are ways of demonstrating the utility of mathematics at the elementary level. With any language a certain distance has to be travelled — a basic framework of grammar and vocabulary has to be built — before the farther, and more exciting horizons come into view. Nevertheless, we could do much more by constructing imaginative exercises and problems (Birdsell 1966; Tenenbaum and Pollard 1963; Bresler 1966; Rashevsky 1968), not only to illustrate the utility and relevance of even simple mathematics, but to enhance the understanding of students by exposing them to very deep concepts early in their geographic careers. Let two examples suffice. The act of *mapping* is familiar to all geographers, but there are not many who see how profound and general this fundamental mathematical idea is. We select, simplify, compress and abstract certain features from the world around us, and so translate them into more understandable and manipulatable forms. We are homomorphic, many-to-one mappers and modellers (Beer 1968), carrying on as part of our traditional inheritance an activity characteristic of many branches of mathematics and science. Even this deep notion is embedded in the still more fundamental one of the *filter* (Hammer 1969), for what does a filter do but select, letting some things pass while retaining others? Like a map, a simple regression or complex partial differential equation is a filter, accepting only some values and rejecting the rest. A trend surface filters the regional signal from the local noise: our spectral analyses filter out the major, linearly additive components of our one- and two-dimensional series; and every component and factor analysis is essentially a decomposing, least-squares filter. Space-smoothing matrices scan our maps, and numerically erode our surfaces by letting only the long wavelengths through; subtract the smoothed version and you have pushed the map through a high pass filter (Tobler 1967; 1969). With such a deep and unifying notion, a map is an equation is a set is a trend is a spectrum is a filter: no need for the singlemindedness of a Gertrude Stein.

But we must face the question of what mathematics we need, and this will depend upon our goals. They are really two: to give the student of mathematical talent and abstract predilection a firm base to build upon later to higher levels of comprehension; and to provide those with other talents the basic linguistic skill that will open the literature by removing fear. We should also recognize with sympathy that for a number of years we shall face a difficult remedial task at the graduate level, and motivation is as important here as for the undergraduate.

Let the integral and summation signs stand for the two major themes in any basic mathematical program for the geographer, indicating respectively the mathematics of continuity and finiteness. In one sense the distinction is not sharp, and many problems can be expressed in either dialect, but some insight into both areas is required for they express different ways of thinking about geographic problems. Many students will require a brush-up of analytical geometry, trigonometry, and coordinate systems, but this can be done with a short course within the geography department using spatial examples, including simple map projections and transformations. Sadly, most geographers cannot handle the basic mathematics of their fundamental tool – the map. The differential and integral calculus should then form the true starting point of the program, if students have not had the elements already in high school (Kruskal 1970).

There are some who argue that in these days of digital computers, where everything is broken down into finite pieces for computing anyway, there is no need for a continuum mathematics with such deterministic overtones for spatial modelling. I have argued this in the past – quite erroneously. But the calculus, if taken just beyond the threshold of ordinary and partial differential equations, is one of the two major skeleton keys of elementary mathematics unlocking many doors to deeper comprehension. It often seems that we have tried to run before we can walk, and the price we have paid for our increased computing power is a severe one. In an age when every high school student can Monte Carlo his diffusion exercise, we have failed to serve a valuable apprenticeship in the tough, and frequently rewarding world of deterministic equations which often stand as the normative rulers against which our spatial realities and simulations may be measured.

On the finite side, the student should acquire a good grounding in the elements of linear algebra and the geometry of vector spaces. The two aspects are complementary and often crucial, for many geographers have quite distinctive spatial minds and require that they "see" the basic operations for fuller understanding. The ability to visualize an algebraic problem in the limited space of three dimensions is well worth cultivating at the elementary level, although it must be admitted that in the long run the algebraicist probably has it over the geometer in the higher reaches of mathematics. In practical terms, however, the advantage of linear algebra comes to the fore when the student blends the probablistic continua of the calculus with the finite pieces of data entering the statistical analysis of a real problem. Linear algebra is the natural mode of expression for all the manipulations involved in actual computing.

The calculus and linear algebra lead the student by parallel paths to statistics, but we should be very careful today to make the first exposure to this important field efficient. Over the past fifteen years we have made many inappropriate borrowings of methods that were devised originally for non-spatial problems at a time when digital computers were unavailable.

Occasionally, geographic experiments can be designed, but there are large areas of important inquiry where inferential notions are not simply inappropriate but totally misplaced (Gould 1970). Often we require good and concise descriptions of large amounts of multi-variate data, and much of a student's time should be spent in handling and *exploring* statistical structures (Klovan 1968). Even distinguished mathematical statisticians are realizing the importance of data exploration in the human sciences as the increased size and availability of computers enlarges our ability to manipulate (Watts 1968).

The ability to use and program a computer is the final fundamental skill that must be required in the mathematical area.

This is virtually a truism today, and most departments require at least two courses in data handling, computer programming and mapping. When these essential empirical skills are added to the solid grounding in mathematics and statistics, the undergraduate major should be able to read the pertinent literature provided in the thematic courses of his junior and senior years, and undertake independent work of substance. For some graduate students, however, the story will be different for a few more years, and there is no question that we should provide imaginative remedial opportunities early in the first year. At this level and age there is frequently strong resentment at being sent straight over to the mathematics department, and the first task is to provide the motivation within the graduate program with imaginative geographic sessions built, perhaps, around the new programmed learning texts (Martin 1969; Bajpai, Calus and Hyslop 1970). With this background at least a minimal awareness will have been generated, and we can expect a considerable increase in communication between theoreticians, empirical workers and practitioners to the mutual benefit and extension of all areas of geographic understanding.

Daydream or possibility?

Is this suggested curriculum, and its supporting parts, a possibility, or is it simply a daydream quite beyond practical hope of realization? The opportunity to write curricula virtually free of the usual constraints comes very rarely, but when such chances do arrive it is worth looking closely at the results. Two programs have been chosen: one at the new University of California at Irvine, the other at the old University of Virginia at Charlottesville.

A student of geography at Irvine (if such is not a misnomer in the strongly interdisciplinary context), works within a school of social sciences devoid of the usual departmental pigeonholes. A typical freshman takes three interdisciplinary courses in social analysis, change and development, and starts a sequence of nine courses in mathematics with work in finite and continuous probability, and the differential and integral calculus (University of California 1970). The other half of his program at this stage is designed to give him breadth by requiring him to elect courses in the schools of biological and physical sciences, fine arts and the humanities. As a sophomore, he is required to take introductions to economics, psychology and sociology, and to continue his mathematical training with linear algebra, differential equations and numerical methods. He also takes digital computing in information and computer science. The rest of his program is completely open for geography courses, perhaps the first being "Model Building in Geography", or "Physical and Man-Made Networks". In the junior year, the mathematical requirement is completed with two advanced courses of the student's choice, advanced work in analysis (such as a Q-technique, psychometrics, experimental games and multidimensional scaling), while in geography, location theory, transportation theory, and problems of urbanization in East Africa might be taken. The senior year is completely open for elective courses, except for a three-term sequence of independent research culminating in a senior project (thesis). Two things are especially important to note: over half of a student's total program is open and elective; and the course offerings after the second year change frequently as the faculty schedules topics that meet their own criteria of interest and relevance. The training of a student, with an emphasis in geography that can exceed the normal major at a traditional university, is rigorous, analytical and tough. As a major innovator and administrator in the social sciences at Irvine once observed to me, "After the third year, I really don't have much more to give the so-and-sos!"

At the University of Virginia several departments, including geography, merged to form a new Program of Environmental Sciences with a heavy emphasis upon ecological systems. The undergraduate follows four major threads in a rigorous, and essentially interdisciplinary program, starting with fluvial processes and air stream modelling. A second sequence engages him in an examination of floral and faunal characteristics and changes, while a third might be fairly labelled applied structural geology and ground water hydrology. These three strands are finally woven with a fourth on urban systems analysis, with a heavy emphasis upon the physical requirements of a city. All majors take a basic sequence in the calculus to ordinary and partial differential equations, together with advanced courses in multivariate statistics, time series analysis and linear programming. More specific techniques are taught within such systematic topics as ecological change and energy budgets. Virtually all courses use the computer extensively for class exercises and projects, and supporting work in social science analysis is available to all students.

It is clear from these examples that much can be done when, for a brief period, the human and organizational constraints are relaxed. Virtually overnight, imaginative extensions can be made that would take decades in departments where the constraints are more binding. And where constraints bind so tightly that a curriculum inches forward with glacial speed, the people who pay the final price are the students. In as much as they have a contribution to make by teaching, research and simply forming a part of a compassionate and well-educated citizenry, the larger society also suffers. In ever-widening areas, it is assumed that a constant process of education is required today. Phrases such as "retreading", "life-time sabbaticals", and "the open university" indicate the willingness to reappraise and reexamine with fresh and open eyes the substance and methods of the past. Nor does such continuing education have to take place within the university. The substance and intellectual excitement of open and evolving geographic curricula should move sooner, rather than later, into the surrounding society.

Beyond the university

Curricula do not stop at university walls, and the new insights and fresh understanding that come from university research and teaching can seep rapidly through an educated citizenry if there are only some who are willing to spend a little time writing for a more general readership. I do not think it is an exaggeration, for example, to label Allen Lane as Britain's greatest educator of the last half century, for his Penguin books opened new worlds of learning to millions. Against tremendous early odds, and in a rather different way, he was following the great Nineteenth Century tradition of Thomas Huxley and Augustus de Morgan (Crowther 1970; de Morgan 1882; L. Huxley 1900; T. Huxley 1870). The same tradition is exemplified in the Soviet Union today, where distinguished mathematicians have written a score of imaginative paperbacks for high school mathematics clubs (Barsov 1964; Vorob'ev 1961). Indeed, it sometimes appears that the more secure a person is, the greater is the willingness to write occasionally for a non-academic audience. Unfortunately, such expository writing is frequently denigrated within the university walls themselves, although it is instructive to see how some distinguished Swedish geographers are willing to write clear expositions of spatial problems of concern to all within the larger society (E.R.U. 1970). Is it possible that people in Sweden are willing to turn to geographers for insight into locational questions and policies of spatial planning because there is a reasonable and contemporary understanding of the subject (Godlund 1961; 1970)? I cannot help contrasting these attitudes with the comment of a distinguished economist in Washington, who was guiding a major program evaluating all the behavioral and social sciences in terms of the contributions they could make to the difficult human problems facing America. In the initial briefing session of the geographers, he opened with the perfectly serious comment "Now let's see, you people deal with topographic maps, or something like that . . . don't you?" The appalled silence which greeted this remark convinced everyone that the job of education beyond the university should begin right there. Two years later, with the final report in hand, he confessed with grace and genuine pleasure that he had no idea "geographers did things like this". But the confession should have come from the geographers: that they had neglected to communicate with the larger society in the past, and so educate the citizens to a larger spatial understanding. In one sense, the open curriculum, where it is possible to achieve it, should be open to all.

Bibliography

ABLER, R., ADAMS, J.S., and GOULD, P. (1971) *Spatial Organization: The Geographer's View of the World*; (Prentice-Hall, Inc., Englewood Cliffs).

BAJPAI, A.C., CALUS, I.M., and HYSLOP, J. (1970) *Ordinary Differential Equations: A Programmed Course for Students of Science and Technology*; (Wiley-Interscience, London).

BARSOV, A.S. (1964) *What is Linear Programming?* (D.C. Heath and Co., Boston).

BARTHOLOMEW, D.J., and BASSETT, E.E. (1971) *Let's Look at the Figures*; (Penguin Books Ltd., Harmondsworth).

BEER, S. (1968) *Management Science*; (Doubleday and Co., Inc., Garden City, New York).

BIRDSELL, J.B. (1966) Some environmental and cultural factors influencing the structuring of Australian aboriginal populations; In Bresler, J. *Human Ecology*, 1966.

BRESLER, J. (1966) *Human Ecology*; (Addison Wesley Co., Reading, Mass.).

BROWN, L.A. and HOLMES, J. (1971) The delimitation of functional regions, nodal regions and hierarchies by functional distance approaches; *Journal of Regional Science* 11, 57-72.

CHISHOLM, M., FREY, A.E., and HAGGETT, P. (Eds.) (1971) *Regional Forecasting*; (Butterworth, London).

COLE, J. (1968-72) *New Ways in Geography: Introduction, Books 1, 2 and 3*; (Basil Blackwell, Oxford).

CRIST, R. (1969) "Geography", *The Professional Geographer*, 21, 305-7.

CROWTHER, J.G. (1970) Augustus de Morgan; *New Scientist and Science Journal* 49, 630-2.

CURRY, L. (1971) Applicability of space-time moving-average forecasting; In Chisholm, Frey and Haggett, (Eds.) *Regional Forecasting*.

DANTZIG, G.B., ORDER, A. and WOLFE, P. (1954) The generalized simplex method for minimizing a linear form under linear inequality constraints, *RAND Memorandum* RM-1264.

DE MORGAN, S.E. (1882) *Memoir of August de Morgan*; (Longmans, Green and Co., London).

E.R.U. (1970) *Urbanisering i Sverige*; (Expertgruppen för Regional Utredningsverksamhet, Stockholm).

GODLUND, S. (1961) *Population, Regional Hospitals, Transport Facilities and Regions: Planning the Location .of Regional Hospitals in Sweden*; (Gleerup, Lund).

GODLUND, S. (1971) Regional and sub-regional population forecasts: Current Swedish practice, in, Chisholm, M., Frey, A.E., and Haggett, P. (Eds.) (1971) *Regional Forecasting* (Butterworths, London).

GOULD, P. (1970) Is *Statistix Inferens* the geographical name for a wild goose?; *Economic Geography* 46, 439-48.

HAMMER, P.C. (Ed.) (1969) *Advances in Mathematical Systems Theory*; (The Pennsylvania State Uni sity Press).

HARVEY, D. (1971) Social justice in spatial systems; *Paper 67th Annual Meeting, Association of American Geographers*.

HUXLEY, L., (Ed.) (1900) *Life and Letters of Thomas Henry Huxley*; (D. Appleton and Co., New York).

HUXLEY, T.H. (1870) *Lay Sermons, Addresses and Reviews*; (D. Appleton and Co., New York).

HÄGERSTRAND, T. (1969) Private communication.

HÄGERSTRAND, T. (1970) What about people in regional science; *Papers of the Regional Science Association* 23, 7-21.
KLOVAN, J.E. (1968) Selection of target areas by factor analysis. *Western Miner* 41, 44-54.
KRUSKAL, W. (Ed.) (1970) *Mathematical Sciences and Social Sciences*; (Prentice-Hall, Inc., Englewood Cliffs).
KUHN, T.S. (1962) *The Structure of Scientific Revolutions*; (University of Chicago Press).
MACKAY, R. (1971) The world of underground ice; *Presidential Address, 67th Annual Meeting of the Association of American Geographers, Boston.*
MARTIN, E.W. (1969) *Mathematics for Decision Making: Vol. 1 Linear Algebra, Vol. 11 Calculus*; (Richard D. Irwin, Inc., Homewood).
MATTHEW, St (circa 50 A.D.) *The Gospel*; Chapter 7, verse 3.
MORISON, E.E. (1966) *Man, Machines, and Modern Times*; (M.I.T. Press, Cambridge, Mass.).
RASHEVSKY, N. (1968) *Looking at History Through Mathematics*; (M.I.T. Press, Cambridge, Mass.).
RICHTER, M. (Ed.) (1970) *Essays in Theory and History: An Approach to the Social Sciences*; (Harvard University Press, Cambridge, Mass.).
SAMUELSON, P.A. (1967) *Economics: An Introductory Analysis*; (McGraw Hill Inc., New York).
SHEN LIN (1965) Computer solutions of the travelling salesman problem; *The Bell System Technical Journal* 44, 2245-69.
SHEN LIN (1970) Heuristic Techniques for solving large combinatorial problems on a computer; *Lecture Notes on Operations Research and Mathematical Systems Theoretical Approaches to Non-Numerical Problem Solving,* (Springer-Verlag, Berlin).
SIMON, H. (1969) *The Sciences of the Artificial*; (M.I.T. Press, Cambridge, Mass.).
SNOW, C.P. (1959) *The Two Cultures and the Scientific Revolution*; (Cambridge University Press).
STEINER, D. (1965) A multivariate statistical approach to climatic regionalization and classification; *Tijdschrift van het Koninklijk Nederlansch Aardrijkskundig Gerootschap Tweede Reeks* 82, 329-47.
STEINER, G. (1969) The retreat from the word; *Language and Silence,* (Penguin Books, Harmondsworth), 31-56.
TAAFFE, E.J. *et al.* (1970) *Geography*; (Prentice-Hall, Inc., Englewood Cliffs).
TENENBAUM, M. and POLLARD, H. (1963) *Ordinary Differential Equations*; (Harper and Row Publishers, New York).
TOBLER, W. (1967) Of maps and matrices; *Journal of Regional Science* 7, 276-80.

284   *Peter R. Gould*

TOBLER, W. (1969) Geographical filters and their inverses; *Geographical Analysis* 1, 234-53.

TRICART, J. (1969) *The Teaching of Geography at University Level*; (George G. Harrap & Co. Ltd., London).

TÖRNQVIST, G. (1963) *Studier i industrilokalisering*; (Geografiska Institutionen vid Stockholms Universitet, Stockholm).

UNIVERSITY OF CALIFORNIA (1970) *University of California, Irvine General Catalogue.*

VOROB'EV, N.N. (1961) *Fibonacci Numbers*; (Blaisdell Publishing Co., New York).

WATTS, D.G. (Ed.) (1968) *The Future of Statistics*; (Academic Press, New York).

WILSON, A. G. (1971) *Entropy in Urban and Regional Modelling*; (Pion, Ltd., London).

ZELINSKY, W. (1971) News from S.E.R.G.E.; Newsletter of *The Socially and Ecologically Responsible Geographer*, May 4, (University Park, Pennsylvania).

13 · Does geography have a structure? Can it be "discovered"? The case of The High School Geography Project*

ROBERT B. McNEE

The challenge

Professional geography in America faced a formidable intellectual challenge in the late 1950's and the decade of the 1960's, one of the most pointed and specific it has ever faced. This came from a broad and pervasive curricular reform movement centered in the sciences and typified by the so-called "new mathematics". One aspect of the challenge was to identify the internal *structure* of geographic thought. As with all growing and developing disciplines, there is great diversity of thought in geography. Yet presumably certain ideas are more central or crucial than others. And, presumably, these central ideas fit together into a coherent framework, a disciplinary structure. Geographers, *as a group*, were challenged clearly to identify this idea-system, to develop new and more effective ways of disseminating it and to put their knowledge of cultural diffusion processes to a practical test by devising new teaching or learning strategies.

How did geographers respond to this challenge? Organizationally, they responded by the creation of the largest professional joint effort at the articulation of geographic ideas that has ever been developed at any time. This was the High School Geography Project, a relatively small project in terms of all curricular reform projects of

* The author wishes to thank all those who responded to a questionnaire about the High School Geography Project, most particularly John Borchert, William Garrison, Hildegard Binder Johnson, Dana Kurfman, John Lounsbury, Alexander Melamid, Waldo Tobler, Philip L. Wagner and Gilbert White.

that time in America but a very large project in terms of previous geographic experience or the size of the profession. Highly innovative learning materials were produced, including particularly the six units of the course, *Geography in an Urban Age* (Macmillan Co) and related materials such as the volume developed to articulate a modern approach to field work in instruction. (Corey et al 1971). Though *Geography in an Urban Age* was developed most specifically for the 10th grade level, so flexible are its strategies that it can be used at the 9th grade and below, as well as upward to the 12th grade and beyond, in to the university system. Judged by its output, the project was misnamed. A more accurate name might have been "The Geography Units Project". Descriptions of these materials abound. (Patton 1970), but in any case, they speak most eloquently for themselves. It is in actual learning situations, in classrooms, that their power to excite, to stimulate and to spark the geographic imagination is most manifest. Geography teachers everywhere owe to themselves, their profession and, above all, their students, to examine these materials with care and to experiment with them. Detailed interpretation of these units here would be redundant.

But the *process* by which geographers moved from an initial perception of the challenge to this ultimate response does need exploration. It is a significant aspect of our intellectual history as a discipline. A very insightful social history of the project has been written already by one of its leading figures, William Pattison (1970). But each of us associated with the project saw it from a somewhat different perspective. This paper stresses aspects of the project experience which seem particularly important *to me*. I hope other social histories of the project will be written, especially by those closely associated with it. Several such histories, taken together, might throw enough light on the project and on our behavior as professionals, so that our effectiveness would be appreciably enhanced "the next time around".

As I see it, the modern world differs from the world of a century or more ago, most particularly in its reliance on *organizations*, bureaucracies if you will. Nowhere are both the advantages and disadvantages of massive organization more evident than in the educational system. The advantages, notably opportunities for specialization and economies-of-scale, suggest that the world of the future will be a world of complex organizational structures, too. The disadvantages of such organizational structures, will also be with us. Most notably, these include tendencies toward lethargy and rigidity. The High School Geography Project was an organizational response by organizational man to problems created by his own characteristics. The High School Geography Project was a *temporary* organization created to overcome some of the rigidities in existing organizations, that is, it was a social instrument. In the future, society will turn again and again to such expedients to keep the great machine from slowing down and foundering on its own rigidities. Therefore, we must learn from this experience. We must see ourselves and our behavior with some clarity, some perspective. I hope this paper will contribute to that end.

A further elaboration of the challenge

On first thought, the challenge to identify geography's *structure* and new means of disseminating it may not seem as formidable as it, in fact, was. After all, geographers have been writing and talking about the "nature of geography" rather loquaciously for a long time. Much soul-searching and ink-spilling has accompanied us every step of the way from an amateurish group of spirited investigators in the 17th and 18th Centuries to the rather more self-conscious professional group of the 20th Century. So interminable has been the discussion, that more than one geographer has confessed to boredom with the topic.

But, in fact, these interminable discussions have generally *not* been in terms of the basic structure of geographic ideas. Instead, they have emphasized the *classification* of knowledge, the place of geography in some overall schema of learning. Most often, the problem of interpreting geographic ideas has been approached as a *territorial* problem. Thus, what are the *boundaries* of geographic thought? What are the boundaries of geography as a whole and of the various subdivisions (subregions?)? What are the "core areas"? In short, the classic approach to the "nature of geography" has been rather analogous to the search for homogeneous regions in regional geography. Such an approach tells us many things about geography but it does not tell us very much about geography's *structure*.

To get at the structure of geographic thought, one needs a stance more analogous either to that of spatial analysis through functional, focal, or nodal regions. (Whittlesey 1954), or to "systems of cities" such as the various central place models. That is to say, interpreting structure requires the identification of nodes, networks and hierarchies of thought within a discipline. Of course, there has to be some concern for disciplinary boundaries, but this can be treated as a problem of communication flows or "distance decay" rather than as a territorial question. By accepting the reality of fuzziness in disciplinary boundaries, one can focus attention on a more interesting intellectual question, the nature of analogies and linkages among geographic ideas.

This is not to say that all discussions about "the nature of geography" have been exclusively classificatory and non-structural in character. Rather, it is to say that the classificatory theme has been so dominant over the structural theme that geographers were relatively unprepared quickly to cope with the challenge to articulate the structure of their discipline.

It is easy to see why this might have been so in terms of the intellectual history of the field. No doubt the long interest in delimiting homogenous regions on the surface of the earth has carried over to the way geographers interpreted their idea-system itself. Similarly, the recent interest of many geographers, including myself, in articulating geography's structure may reflect an adherence to "Haggettry". But I think the territorial approach to the nature of geography also reflects the bureaucratic world in which most of us function. Modern "disciplines" in part reflect nodalities in thought but they also reflect administrative convenience within universities. In the Byzantine world of the modern university, one is often called upon to "justify" the "territory occupied" by a department such as geography. The "boundaries" between disciplines may be of limited import in the realms of "pure thought" but they are of great practical significance in university administration, particularly at budget time. University men quickly learn of the necessity to be constantly on the alert to defend the territorial "borders" of their discipline. Why should we expect that a more purely intellectual issue, the structure of geographic thought, would engage geographers more deeply and more consistently than these practical issues of survival and growth related to the social environment, the bureaucratic world of the university? In short, the call to articulate the structure of geographic thought challenged geographers to break with long-established patterns of thought which stressed classification which had been buttressed by both the intellectual history of the field and the social environment in which most geographers function.

The other major thrust of the curricular reform movement, the enthusiasm for developing new learning strategies, was also a greater challenge to geography that it might at first appear. In stressing disciplinary structure, the curricular reform movement was expressing one of the most basic values of science, *the love of order*. But science has other values, too. One of the strongest is a' *love of inquiry* or exploration. In the curricular reform movement this was widely expressed as an advocacy of "discovery learning". The reformers generally believed that a love of inquiry had led them personally to discover the order of the universe articulated by their discipline and that this principle should be made the basis of instruction in the sciences. Thus, they believed that a carefully structured sequence of learning experiences stressing "discovery" would lead the student to a gradual understanding of a discipline, with much "joy of learning" in the process. It was generally assumed that the twin values of order and inquiry were complementary rather than conflicting values.

In some respects, geography was better prepared to meet such a challenge than many disciplines. Geography teachers have long stressed "discovery learning" without using that term to describe their teaching strategies. For example, why have geographers used topographic sheets, air photos and field trips in instruction as much as they have if not to attempt to replicate "discovery" experiences that they themselves have known? For generations, geography teachers have known, intuitively, that what the student manages to discover for himself is more deeply "learned" than that which is presented in prepackaged, lecture, form. All they have lacked has been an "objective" explanation for this psychological phenomenon.

Yet it must also be admitted that the "discovery" approach, under whatever name, has not been the dominant tradition in American geographic education. Rather, the dominant tradition has been much more didactic. It is only here and there that the "lecture-textbook-fill in the blanks" syndrome has been seriously challenged. Perhaps, in part, this has been because so many of us want to "cover" all aspects of geography and have lacked a widely accepted description of the structure of geography, leaving us with endless territories to "cover".

But even if the "discovery" tradition had been stronger than it was in geography, the changes in the field in the last few decades had outrun the development of "discovery" learning strategies. Field trips, topographic maps and air photos could lead one into certain aspects of geography. But what about the behavioral thrust in geography, the stress on the complexities in geographic decision making? Some new kind of discovery experience would be essential if such ideas were to be made real for the student. Or, what about the deductive element in the ideas of von Thunen, Weber, Christaller, Loesch, and the like? Such ideas were becoming very important in geography. Discovery strategies that were appropriate for an inductive, empirical, field were less applicable as geography became more attuned to models and other forms of deductive thinking. Or, the ideas relating to cultural diffusion? Timeworn discovery strategies of a heavily *local* nature (such as topographic maps, air photos or field trips) needed rethinking in the diffusion context. And so on.

In short, the challenge to develop new teaching strategies emphasizing discovery required geographers to de-emphasize their dominant pedagogic tradition and to highlight their lesser pedagogic tradition of discovery. These strategies had to be applied to geographic thought as it was in the latter half of the 20th Century, not just to what it had been in the past. So the challenge was a formidable one indeed.

The unanswered question in the late 1950's was whether or not American geographers would adequately perceive the challenge of the curricular reform movement and, if they did perceive the challenge, would they devote equal attention to both its aspects? Or would they stress either the identification of structure or the development of learning strategies more than the other? Before considering such questions here, I want to explain how I view the challengers, the curricular reformers outside the field of geography itself.

The challengers

That the curricular reform movement of the late 1950's and the decade of the 1960's in America came to exist at all or that it took the form that it did is inexplicable except in terms of some assumptions about American culture. In interpreting American culture, and therefore its educational behavior, I think there is value in assuming a polarity of conflicting values: the vigorous pursuit of excellence on the one hand and a great longing for equality on the other (Gardner 1961). Americans believe in both values, though in varying degrees, and actual educational programs tend to be compromises between them.

In the late 1950's, this tension expressed itself as a cry for a greater emphasis on excellence. No doubt competition with the U.S.S.R. partly accounts for this. The launching of Sputnik produced much sputtering in America. But, perhaps it was more that those Americans most attracted to the dream of excellence saw in the reaction to Sputnik a lever with which to move American education and hence American culture toward the pole they favored. They saw a lever — and used it. A cry for "reform" went up. The educators in the schools were said to have "failed" because they had not stressed excellence enough.

If the educators had "failed", where was one to find the competence to develop a curriculum with the proper emphasis on excellence? Obviously, one should look to those who had demonstrated their devotion to excellence through their own research, to the men of science. *Their* devotion to excellence was untainted by past compromising with other social values. And how were these men of excellence to be identified? Through the disciplinary professional societies, of course. Heretofore, such societies had concerned themselves primarily with the creation of knowledge and with only limited aspects of the dissemination of such knowledge. Dissemination among the professionals of that discipline had been of central concern but wider dissemination of only peripheral interest. Now these societies were to broaden their view so as to help society as a whole to move toward excellence.

This was an attractive argument for many scientists and men of unquestioned excellence in many sciences were associated with the

national curricular reform movement. Because they were so inti-
mately associated with the actual growth of new ideas in science,
such men were very aware of the rapid obsolescence of scientific
ideas. They were discouraged by the predominance of archaic ideas
in the 'science' being taught in the schools. Hence they stressed
economy in educational effort. That is, they stressed the need for a
rigorous sifting of the ideas of any given discipline to sort out which
were most crucial in understanding the discipline as a whole. In
short, they believed that an interpretation of a discipline's *structure*
should precede the development of learning strategies.

These scientists also knew that they themselves had been led on
and on into the heart of their discipline by the joy of solving
intellectual puzzles. They were appalled at the emphasis on
memorizing 'facts' in most 'science' courses. If "discovery" had been
so important to themselves in learning the structure of their fields,
why shouldn't it be equally important to everyone?

All such a train of thought lacked among the men of excellence in
the curricular reform movement was an articulate spokesman around
whom they could rally. The movement found such a hero in Jerome
Bruner, a psychologist with great interest in the processes of learning
(Bruner 1963). It seems to me that Bruner's primary contribution
was to provide a psychological rationale for what the men of science
wanted to do anyway. Not all of the curricular reform projects of
the period bore his stamp, but most did.

In retrospect, one can question some of the basic assumptions in
this line of thought. One of the basic difficulties with science is that
it must be pursued by human beings. Hence science, in the end, rests
on assumptions or beliefs just as do other forms of knowing. An
implicit assumption in Brunerian thought is an idea held by most
scientists, that science moves from where it is toward some goal set
by nature in advance. In this view, the changes in a discipline over
time are seen as a progressive unfolding of "the truth" about the
universe. If one did not believe this, how could one assume that
"discovery" experiences would necessarily lead to the understanding
of a particular disciplinary structure? But Thomas Kuhn (1962)
argues that science may *not* move toward some goal set by nature in
advance. Rather, that science may, in fact, have no external goal. It
may advance, instead, from what it knows toward what it *wishes* to

know. In short, the disciplines are primarily a social phenomenon and only partially a reflection of any "truth" that may exist in nature. If Kuhn's ideas are accepted, then there is real reason to question whether or not the historically and socially derived *structures* of the disciplines could ever be comprehended through true "discovery" experiences. In education, one would have to choose between stressing "discovery" which might well lead toward unique thought structures rather than those espoused by the existing disciplines. Or stressing disciplinary structures which might be learned through pseudo-discovery experiences but which would be subverted if true "discovery" had full expression. But such questioning, that I or others may now engage in, had little significance in the initiation of most of the curricular reform projects. In the curricular reform movement there was a general confidence that the goals of order (disciplinary structure) and inquiry ("discovery") were complementary rather than divergent.

The response of geographers to the challenge: The creation of the High School Geography Project

Probably some geographers perceived the general curricular reform movement as a threat, others as an opportunity. Some saw that if geography did not join the curricular reform movement and produce new materials of excellence, it might well lose even the precariously low "status" position it had attained in America. The "image' of geography could not stand much erosion, since geography was already well down on the "pecking order" listing of the disciplines, though by no means at the bottom. But the general curricular fluidity in the schools created by the demand for excellence of the curricular reform movement could be interpreted as a great opportunity. This fluidity gave the professional geographers their greatest opportunity in many decades to present the best of geography to a wide audience. Through involving themselves in curricular reform, geographers might permanently alter the public "image" of school geography. They might be able to demonstrate that geographers not only dealt with serious intellectual questions but also that they were capable of presenting them imaginatively.

What geography needed to make such a possibility into a reality was someone with the imagination and drive of the general leader of the reform movement, Jerome Bruner. Gilbert White was such a man. White's research had established him as meriting the title of "man of excellence". Further, he had had much experience as an administrator, as a leader of men in a bureaucratic age. As a college president, a Washington bureaucrat, and a member of one of the oldest and most prestigous geography departments in the United States (University of Chicago), White had the background to launch a curricular reform project of the dimensions required. Either it must be massive, in terms of previous organized efforts by geographers, or it would be ineffective.

More than most geographers, in America, Gilbert White was much concerned with questions of public policy. In his view, geography had much to offer in helping to resolve some of the great social questions of the age, such as world peace and environmental management. Therefore, he favored the enhancement of geography in the schools as much or more than anyone, but more from this public policy stance than from a narrow position of disciplinary striving. He made the launching of a geographic curricular reform project one of the major aims of his administration as President of the Association of American Geographers (1960-1).

Of course, other geographers were also alert to the opportunities for geography in the reform movement, so they joined eagerly in the effort. Together, the Association of American Geographers and the National Council for Geographic Education (the professional society long concentrating on such questions) formed a Joint Committee on Education to develop and implement a plan. Gilbert White and Clyde Kohn became co-chairman. Clyde Kohn had made many contributions in urban, settlement and social geography and therefore met the "man of excellence" criterion. But, in addition, he had worked long and hard in the field of geographic education over the years, whether it happened to be a "popular" topic or not. By 1961, the Fund for the Advancement of Education (the Ford Foundation) had begun funding the project, though only at a modest, experimental, level. William Pattison became Director.

The project held a series of meetings designed to pinpoint just what the major ideas of contemporary geography as understood in America were. The ideas identified (Joint Committee on Education

... 1962) became basic raw material for an "experimental year" during which "men of excellence" (i.e., selected academic geographers) were teamed with classroom teachers from the schools released from other duties for such experimentation. Some of the more interesting of the teaching units developed were subsequently published (Kohn 1964).

Conceptually, the discussions among geographers as to the contemporary nature of their field spawned one of the more significant philosophical papers ever produced in geography. This was a paper by William Pattison (1964), on the diverse conceptual traditions of the field. As Director of the project, Pattison was at the center of a rather large communications network among geographers. The messages on this network were diverse, to put it mildly. In particular, some geographers accorded a central position to the idea of "areal association" (or "areal differentiation"), an idea well articulated by Richard Hartshorne and others before him, such as Alfred Hettner. In the 1930's and the 1940's, this had been the "establishment" view in American geography. But in the 1950's, this view had been challenged by the "new geography" generally flying under the colors of "spatial analysis". Actually, the renewed emphasis on "spatial analysis" was not as "new" as it sometimes sounded. A careful reading of Hartshorne, Dickinson (1969), and other pundits, shows that this emphasis has seldom been absent from geographic research in the last century, though more prominent in the work of geographers such as Friedrich Ratzel or Sten de Geer than others.

In the 1950's, most of the sound and the fury of intellectual debate had to do with the attempt of advocates of this suppressed spatial tradition to upset the central position held by the "areal differentiation" tradition. Yet, it was clear to Pattison that there were at least two other major threads or themes in the general geographic tradition. These, too, had merit and should be recognized, even if the debates of the moment did not revolve around them and their claims. These were the "physical geography for its own sake" tradition and what might be called the "environmental", "ecological", or "man-land" tradition.

The historic method of dealing with the waxing and waning of these diverse traditions had been rather authoritarian. That is, at any one point in time, consensus among American geographers had been achieved by declaring one particular tradition or pair of traditions as the "one true faith". Other traditions were relegated to subsidiary status or labeled as "deviations" (heresies?). William Pattison proposed a more ecumenical approach. He saw nothing particularly threatening in accepting the diversity of geographic tradition and going on from there. The intellectual challenge for geographic education was to discover how to relate the best of all traditions rather than how to suppress some and highlight others. Pattison's paper was a germinal one, both directly with regard to the Project, and also more widely.

Having learned many things about geography and about geographic education from the experimental year, the project sought, and gained, funding on a larger and more long-term basis from the National Science Foundation (1964-1970). Concurrently, the Commission on College Geography of the Association of American Geographers sought and obtained support from the National Science Foundation. No doubt the two requests were mutually supportive. Henceforth, the histories of the two projects were partially intertwined, though the projects differed not only in the educational levels at which they focused but also in their strategies. In general, the High School Geography Project followed a "rifle" approach, stressing the development of integrated materials for a particular grade level (the 10th grade) and particularly in the "social studies" context, whereas the College Commission followed a more "shotgun" approach, producing a great variety of materials rapidly on diverse and only partly related topics. Nevertheless, there was much feedback of ideas between the two projects, in part because of careful efforts at liaison and in part because of double involvement in educational reform by some geographers.

Support by the National Science Foundation had significant psychological effects, quite apart from the generous funding itself. Such support implied to many geographers that geography had won "acceptance". It tended to direct attention away from a narrow concern for "status" and toward more significant questions such as the rigorous development of geographic ideas. In short, it released constructive energies in geographic education.

Support by the National Science Foundation also brought a shift in responsibility for the project. Responsibility was now concentrated in one professional society rather than two. And, of course, the one chosen was the one whose values most clearly reflected those of the reform movement as a whole, particularly a dedication to "excellence" in substantive research. This was the Association of American Geographers.

In turn, the Association of American Geographers centered responsibility in a new Steering Committee. Wisely, the Association encouraged the Steering Committee to function with considerable autonomy and boldness of vision. Gilbert White was named Chairman and the other members of the Steering Committee were rotated during the life of the project, so that a large number of well-known American geographers served on it at one time or another. (Pattison 1970, 76). However, Gilbert White continued as Chairman throughout, providing continuity and a driving vision. The ultimate success of the project is a monument to his perseverance and dedication, and to his persuasiveness in enlisting so many able geographers in the furtherance of a noble dream.

The national prestige of the general curricular reform movement, the aura of "excellence" embodied in that movement, sponsorship by the Association of American Geographers, and Gilbert White's skill as a persuader induced many of the leaders of American Geography to serve on the Steering Committee. The early years of this Steering Committee (1963-1965) were most crucial in setting directions and therefore are of the most general interest. Later, as the basic decisions of those years were implemented, the questions at issue became increasingly technical in nature. For example, after 1965, much time was devoted to choosing and implementing appropriate strategies for trial of the units produced. These latter years are of much interest to those who are students of organizational behavior, such as myself. At its height, the High School Geography Project was a rather large enterprise, employing many people in whole or in part, scattered over much of the North American continent, though focusing at the Boulder headquarters. In attempting to coordinate the behavior of so many diverse people at diverse locations in an atmosphere of creative freedom, the project faced many of the dilemmas of internal communication, awareness of environment, and feedback that are faced on a more permanent basis by governmental bureaus or corporations. No doubt a full understanding of the "spatial interaction" within the project itself would clarify many questions we now have about its final output. And, no doubt if some geographer had sought to study the spatial interaction of the project in process, we might know much more than we now do about how all organizations function in space. But not every geographer is as interested as I am in the geography of organizations. So the following discussion focuses on only two questions: (1) the response of the project to the challenge to articulate geography's *structure* and (2) the response to the challenge to develop "discovery learning" materials.

The response to the challenge to articulate structure

How did the High School Geography Project respond to the specific challenge from the general curricular reform movement to articulate its *structure*? Hypothetically, the project might have produced a comprehensive and fully articulated description of the internal network of geographic ideas in the sixties that would have had rather widespread effects on geographic instruction at all levels and, indeed, on some aspects of future research. In science, there is always a relationship between the order perceived in existing knowledge and the additional knowledge which is sought through research. But the project did not do so. It is a moot point as to whether it *should have* or not. The important point here is that it did not. This is not to say that the course materials produced, *Geography in an Urban Age*, do not reflect the structure of geography as perceived in the 1950's and 1960's in America. Indeed, one can probably learn more about such perceived structure from the materials than is generally the case in High School texts. Yet the overall emphasis in the course produced is on imaginative ways to learn about particular geographic *topics* (cities, culture, political processes, etc.). The overall emphasis is *not* on how certain key ideas in geography relate to each other and together produce a comprehensive system of thought.

Had there been great emphasis on geography's structure, then each unit would have built upon the other in such a way that the last unit might well have been a "capstone" unit in which the students themselves articulated the overall system of thought involved and its implications for further study or research. Instead, one of the virtues of *Geography in an Urban Age* is its great flexibility. The course can be used as a whole, in the suggested sequence. Or, the whole course can be used, but with the units in altered sequence. Or, individual units may be used with conventional geography courses or with social studies courses. In this great flexibility, the course is very much in the geographic tradition. Distinct "topical" or "regional" units, which can be shuffled more or less at will, is a very common characteristic of geography courses at all levels. In short, the High School Geography Project responded much more fully and completely to the challenge to produce "discovery learning" materials than to the challenge to set such materials within a clearly articulated structure of geography.

This does not mean that there were no attempts in the project to articulate such a structure. On the contrary, there was much concern with a structure in the earlier phases of the project. Some of the more interesting aspects of the struggle to articulate a structure will be discussed here, not only because they form an important aspect of the project's history, but also because it seems evident that geographers will be struggling with such a pin-pointing of geography's internal logic for a long time to come. William Pattison's paper on the four traditions of geography and the Advisory Paper were early attempts in the project to identify structure. Later, Edwin Thomas (1964) developed a way to interpret geography's structure according to the ideas of the spatial analysis point of view. However, the greatest effort toward specifying geography's structure occurred during the "age of the outlines" in the Steering Committee in 1964-65. During this period, the Steering Committee became, in effect, "an assembly of philosophers", as Pattison (1970, 63) has noted. Actually, the "circle of philosophers" noted by Pattison was two circles interacting, since the Geography in Liberal Education Committee (predecessor of the Commission on College Geography) was tackling similar questions at about the same time. (Association of American Geographers 1965). Some of the "philosophers" in question were involved in both groups.

The philosophic discussion in the Steering Committee was very broad ranging and sometimes heated. The stakes were high. The course (or courses) to be produced by the project presumably needed some overall design or framework. But which design? The designs proposed reflected deeply held convictions about how geography was (or *should be*) structured. The Steering Committee was probably as representative a group of geographers as had ever been assembled in America. Hence there was a healthy diversity of outlook. In favoring a "world regional approach", some geographers seemed to be giving primacy to the concept of planetary unity, second place to the concept of culture regions, and third place to other geographic concepts. Others similarly stressed planetary unity but gave second place to the various aspects of the "physical base", third place to man's use of the earth, and fourth place to man's cultural differentiation. Others favored a developmental approach, elaborating the structure of geographic thought by recapitulating its historic evolution, beginning with the kinds of geographic questions raised and answered in one way or another by the ancient Greeks. It seems likely that structural ideas about geography that would loom large in such a course would include planetary unity, exploration, mapping and regionalization. Several suggested outlines were strongly *process* oriented, reflecting the research trends of recent decades; of course, these outlines differed in just which sets of processes were most stressed – climatological, urban, economic, political, cultural, or what-have-you. Decision-making and human behavior were strongly stressed in at least two outlines, reflecting not only a research trend of the 1950's but of the 1960's and early 1970's as well. Thus, there was no lack of variety in viewpoint as to what geography's structure might be. One outline, taking a world social problems approach, met with little favor, in part, at least, because of its lack of emphasis on structure. Today, when disciplinary structure is receiving less attention and "social relevance" more attention it might well be that a course stressing such social problems as population growth, economic development, or environmental quality might command everyone's attention. But, not so in the early 1960's. Then the stress was on structure. Among all of the diverse alternative paths (course outlines), a choice had to be made. Such a choice hinged on some measure of consensus as to what the structure of geography was or ought to be.

In December, 1964, the Steering Committee chose the "McNee Outline", later to become known as the "Settlement Theme

Outline", to guide course development, for 1965 at least. More specifically, the Steering Committee chose "Part A, Topical units", of that outline plus unit 15, "Geography as a structured discipline" for further elaboration. It recognized that "Part B, Regional units: the great culture regions" could, in itself, represent a full year's course work. For the time being this, or any other, regional approach was to be set aside and effort concentrated on the topical and structural portion of the outline. As it turned out in the end, the hypothetical regional course was never developed and the only specific bow to comprehensive regional analysis made by the project was the unit on Japan, added somewhat later to the McNee design. This outline of December, 1964, was subsequently elaborated in a series of about 8 drafts, culminating in a still provisional draft of May, 1966, published in limited edition as the Settlement Theme Course Outline. By that time, the initial emphasis on structure had already begun to fade somewhat in favor of an increasing emphasis on "discovery learning." But, for the historical record, the outline of December, 1964, is here reproduced in its entirety:

Suggested outline for a tenth grade world geography course

by R. B. McNEE

A. TOPICAL UNITS

Unit 1: Intra-city analysis

Observation and interpretation of the internal patterns of cities. The city as both a place of work (production system) and as a place to live (consumption system). Hypothetical city plus a local case study.

Unit 2: Inter-city analysis

Central place functions. The classification of cities. Individual cities related to the general settlement network, including regional hierarchies.

Unit 3: Manufacturing and mining as settlement-forming activities

Basic location theory, stressing settlement-forming activities rather than settlement-serving activities, with both considered as parts of a settlement matrix.

Unit 4: Primary production activities as settlement-forming activities (Mining treated in unit 3)

Idealized interpretations of the settlement-forming aspects of gathering, herding, fishing, forestry, and agriculture. Heavily modified Von Thunen model and other historic settlement patterns. Primary stress on agriculture and forestry, with the development of commercialized agricultural regions and commercialized forestry regions considered as a function of the growth of urban-industrial settlement systems.

Unit 5: The culture base

Culture hearth, culture region, origins, dispersion, diffusion. Cultural geography presented as a system of constraints within which the settlement processes of units 1—4 must operate. Similarities and differences among the major culture regions of the earth in terms of the cultural framework for settlement systems.

Unit 6: The resource base

Physical geography and biogeography presented as a system of constraints within which the settlement processes discussed in units 1—5 must operate. Changing resource opportunities, barriers, and problems in relation to changes in the settlement processes discussed in units 1—5.

Unit 7: Detailed consideration of one resource complex

Water resources unit?

Unit 8: Territorial states and nation states

"Countries" as spatially ordered systems which can be analyzed geographically in terms of culture base, resource base, and settlement system(s). The state as one of the most complex and fragile of geographic systems. Comparative analysis of states.

B. Regional units: the great culture regions

Each culture region considered as a laboratory in which to test and elaborate the ideas developed in units 1—8. Each region treated in same topical sequence, to allow ready comparisons among regions. Preferred sequence: culture base, states, resource base, and settlement system(s). Settlement analysis to include intra-city analysis of at least one major city in each region.

Unit 9: Western European culture region

North Sea core area outward to Iberia, Sicily, Austria, East Germany, and Scandinavia. Emphasis on the evolution of settlement-system since 1750 (*NOT* on earlier developments).

Unit 10: Overseas European culture region

United States, Canada, Australia, New Zealand. The area outside West Europe in which West European culture and settlement system has been most fully developed and elaborated.

Unit 11: Eastern European culture region

Slavic heartland outward to Slovakia, Bulgaria, Siberia. Area heavily influenced by outward movement of West European culture but in which this culture complex has been heavily modified. Urbanization and related "westernization".

Unit 12: Latin American culture zone

A zone of culture conflict: remnants of Indian culture, "Latin" culture deriving from fringe of West European culture hearth, pressures from both U.S. and West Europe, etc. Problems of territorial state organization. Problems of the resource base. Fragmentary settlement systems. Rapid growth of primate cities.

Unit 13: Asian culture regions

East Asian culture region (Hwang-Yangtse plain outward to Manchuria, Sinkiang, Tibet, Formosa, and Japan) and South Asian culture region (Ganges Plain outward to Burma, Ceylon, W. Pakistan) and zone of conflict (Southeast Asia).

Unit 14: Middle East and Africa

Southwest Asia and Egypt outward to Morocco, Somalia, W. Pakistan, Mongolia; Trans-Sahara Africa as zone of culture conflict.

Unit 15: Geography as a structured discipline

Formal and explicit interpretation of geography as an ordered, disciplined, way of interpreting reality, based on ideas first introduced in units 1—8 and subsequently presented in more detail in units 9—14. The big ideas of geography. The implications of these ideas for further study.

Why was this particular outline chosen over its competitors? No doubt there were many reasons, but I think the most significant single reason was that it spoke more directly to the demand for *structure* than most of its competitors. The demand for structure was strong in the early 1960's. It came not only from the general curricular reform movement but also from certain segments of the profession itself, particularly from many associated with the "new" geography. The new geography tended to put great stress on order, rationality, logic, deduction and model building; it tended to downgrade empiricism except when clearly in the service of rational analysis. It stressed the identification of simple, clear, concepts (such as distance, direction, velocity, or scale) whose inter-relationships could be expressed in clear ways, hopefully in the most un-ambiguous language of all, mathematics.

My assertion of the importance of *structure* as an attractive feature of the "McNee Outline" is supported by the subsequent elaboration of the outline in various drafts. For example, the draft of January, 1965, stressed that each unit of the course should "consistently develop not only specific concepts but also analogies among concepts". It held that geographic concepts could, in general, be grouped under three headings or "concept clusters" and that these were *"areal association* (associations in a place), *spatial interaction* (associations among places), and the *spatial and time scale modifiers* (the effect of both spatial scale shifts and time scale shifts on the interpretation of both areal association and spatial interaction)". It was assumed in this draft that more specific concepts appropriate to these headings would be developed in each unit and that, unit by unit, analogies among such concepts would be developed. For example, in the intra-city unit, the general concept of areal association might be developed through such lesser concepts as site, form, or homogeneous regions. Similarly, spatial interaction concepts could be developed through central place, range, threshold, linear accessibility, accessibility on a network, or nodal regions. Or, skipping to unit 4, primary production, one would find areal association represented through site factors or homogeneous agricultural regions and spatial interaction represented by such concepts as complementarity, regional specialization, comparative advantage, and variations in cropping intensity (von Thunen rings). Or, in the culture unit, areal association would be expressed through culture hearths, culture regions, or cultural core areas whereas spatial interaction would be expressed through diffusion and related concepts. It was assumed in the outline that, by such a consistent stress on analogies among concepts used in geography in treating different classes of phenomena (topics such as cities, nation-states, habitats, and so on), that the overall unity of geographic thought would be evident to the student by the last unit of the course. At that time, such analogies could be further clarified and the student would indeed emerge with a clear view of "the geographer's way", the structure of geographic thought. Of course, this elaborated outline carried many suggestions for skills development appropriate to such concept development.

I suppose that a few very specific aspects of this particular scheme for elaborating geography's structure were uniquely my own. However, in the main, it reflected such consensus about geography's structure as existed at that time in the United States. It was based heavily on all preceding work in the High School Geography Project, on parallel discussions in the Commission on College Geography, and on much interaction within the Steering Committee. In particular, the revised outline reflected concepts in the other outline proposed which most stressed structure (Gilbert White's outline). Successive revisions of the outline reflected other structural ideas from members of the Steering Committee.

No doubt there were other lesser reasons, besides its stress on structure, for the acceptability of the "McNee Outline". For example, the idea of starting with the city and building outward was quasi-novel; the Steering Committee was under considerable psychological pressure to show at least some interest in novelty. Similarly, the outline left room for all traditions in human geography to find some expression, even though it gave "pride of place" to the spatial analysis tradition. Further, the outline strongly reflected the research trends of the preceding decade. The "new geography" was not only quantitative; it tended to be topical (rather than regional), process-oriented, reflective of the ideas of systems analysis, urban-oriented, and integrative or convergent in terms of historically separate research paths in geography (in particular, in the 1950's and 1960's, there was a tendency for certain aspects of cultural geography such as diffusion studies to converge with urban and economic geography). All of these aspects of the research trends of the time were reflected in the outline or in the oral elaboration which I and others on the Steering Committee brought to its discussion.

But if it was primarily the emphasis on structure that made the outline acceptable to the Steering Committee, then one must ask why the final product in 1970, *Geography in an Urban Age*, does not reflect more of this early concern for structure? I think the most important reason was the sheer immensity of the task of defining the structure of geography. The need for a clarification of geography's structure was felt so strongly precisely because so much difference was perceived between the dream (a tightly organized and rational body of knowledge) and perceived reality (the immense variety in what passes under the label of geography). In this particular context, high school materials, the effort may have been premature. But for the field as a whole it was not premature. I believe that the search for structure is a road down which geography must travel if it is to be taken as seriously as it ought to be by men of learning. It is evident that many geographers today share this view that a search for logical coherence in geographic thought is imperative. Two significant attempts to move down this road are the recent monograph by David Harvey (1969) and the new text by Ronald Abler, John Adams and Peter Gould (1971).

In moving down this road, I think the approach taken in the High School Geography Project is the one likely to be effective in the long run. That is, I think that perceiving the structure that already exists in geography, or is emerging, should be the point of departure. Such a structure probably includes a great many unconscious stances that are there in our minds, if we just ferret them out. The role of values and taboos in our thinking particularly needs exploration. That is, I adhere to a social interpretation of geography and believe that its structure is very much a product of its social past. Among other things, this past includes a preoccupation with occidental modes of thought (as opposed to more universalistic modes of thought), world conquest and imperialism by the occidental powers, a preoccupation with property (and hence, physical objects), a progressive alienation from the physical and biological world expressed in exaggerated ideas about an abstraction called "Nature", a general lack of familiarity with one of the greatest philosophical revolutions of our time, that sparked by Sigmund Freud, a premodern preoccupation with individual decision-making in a world increasingly patterned by the decisions of massive organizations, and a general uneasiness with the whole topic of social values. It is as we truly understand our own cultural past as geographers that we will be able to give up some of the understandings of geography's structure that we now have and move on to a conception of geography's structure more in tune with today and tomorrow. Ideas from non-practitioners of geography, such as the philosophers of science, should be used wherever practical in interpreting our own thought-structures. But the aim should be to ferret out what the structure of geography *is* and *is becoming*, not what it *ought to be* according to some philosophic schema imposed from without.

Another probable reason for the gradual fading of an emphasis on the identification of structure in geographic thought within the project has already been suggested in an earlier portion of this paper. It may well be that the two values assumed to be complementary in the Brunerian model for educational reform were in fact divergent values. Perhaps there is always a strong tension between the value of order (represented by a stress on structure) and the value of inquiry (represented by a stress on discovery learning).

The initial emphasis on structure in the project *could* have been maintained if that had been the dominant value of the members of the Steering Committee, the Director, the Unit Authors, and the central staff in Boulder. For example, it could have been maintained if the output had been conceived as a textbook (rather than the multi-media course it turned out to be) and if such a textbook had been produced at a series of writing conferences at a central place at which the differing ideas about the structure of geography could have been threshed out over and over again. Such a highly structured text would have stood as a monument, no doubt an impressive monument, to such consensus about the structure of geography as existed at that time. It would have been a structural milestone against which future changes in the idea-systems could have been plotted. It would have clarified many things about geography for geographers, other professionals in other disciplines, school administrators, school teachers, and many, many students.

But would such a structural text have excited more students about the potential that exists in their own minds for imaginative exploration of their geographic world? Would it have done more in the long run to make geographic thinking a deeply experienced, and therefore real, part of the student's mind? Of course one can seldom be sure. But I think not. I think the course actually developed does rather more to free and excite the minds of the students.

The response to the challenge to develop discovery learning experiences

Few, if any, of the various curricular reform projects of the 1960's responded more fully or more imaginatively to the challenge from the curricular reform movement to develop "discovery learning" strategies. As a relatively small discipline, geography is often cast in the role of "follower", both in research and in instruction. But, in this case, geography was clearly a leader. No doubt many aspects of the six units of *Geography in an Urban Age* will be emulated by other disciplines, as well as by geographers developing other learning materials. (Hill 1970).

That there were problems in pursuing the initial structural emphasis seems clear enough. But that does not explain the great success in pursuing the other theme, discovery learning. Certainly one reason was the decision by the Steering Committee to allow decentralized unit authorships. The Committee wanted to interest the very best talent available in developing the units. To do this, the Committee had to be flexible, to allow such men to maintain residence at their various university posts across the nation. The unit authors selected tended to be relatively young men, already well launched on their careers, but not yet frozen into particular stances in relation to the discipline. Selection as a unit author often represented a great professional opportunity for such men. They had a chance, through their unit, to project themselves nationally. They had little to lose and much to gain by being as innovative as possible. Being innovative in teaching strategy was more viable than trying to "out-philosophize the philosophers" on the Steering Committee with their preoccupation with structure. Then, too, though all units were exhaustively tested and retested in classroom situations, this evaluation was always by unit and never for the course as a whole. This piecemeal approach to evaluation was necessary in order to meet production schedules. Had a more comprehensive evaluation been possible, there could have been tests of the *transferability* of concepts from one topical setting (for example, the urban unit) to another topical setting (for example, the culture unit). In short, a comprehensive evaluation would have strengthened the structural goal whereas the unit by unit evaluation adopted strengthened innovativeness within each unit.

Another reason for the general enthusiasm for discovery learning in the project was familiarity of some members of the Steering Committee with Brunerian thought. Gilbert White had met Jerome Bruner personally and others on the committee, including myself, were familiar with his published work. Some of the very first units initiated, particularly those of Arthur Getis and Howard Stafford, stressed games and related forms of discovery learning. These were well received, both in the classroom and outside it (for example, on the Steering Committee). Thus a "climate" in favor of discovery learning was established early in the process. This was buttressed by the inputs of psychologists such as Leonard Lansky. No doubt there were still other reasons for the great emphasis on discovery learning in the project. But perhaps the most important single reason of all is that the Project Director, Nicholas Helburn, developed a deep knowledge about such learning strategies and became a tireless advocate of them.

But to say that the project emphasizes "discovery learning" is only one aspect of the question. In fact, there are many, many ways to discover things and though *Geography in an Urban Age* uses a wide variety indeed, it is quite remarkable how often a "gaming" or "simulation" kind of discovery strategy is used. There are games involving the growth of a city over time, locating a plant, farming, the spread of ideas, political decision-making, mining and land use, and so on. Indeed, one of the more interesting by-products of the project was the preparation of a separate pamphlet on the possible use of simulation generally in instruction (High School Geography Project 1970). This illustrates the generalized learning that occurred in the project about geographic education, learning which affected many members of the profession quite deeply and therefore may have some rather significant effects on geography in general, quite apart from the specific materials produced in the project.

The great emphasis on gaming and other kinds of discovery learning was rooted in great concern for learning as behavior, for the behavior of the student in a learning situation. In developing this way, the project was moving in directions rather parallel to the field as a whole. A behavioral approach is much more common in the geographic literature today than it was in 1961. Whereas the decade from 1955 to 1965 was a period during which many relevant ideas from economics were being absorbed by geography, the post-1965 period has involved less reliance on rationalistic, calculating, all-knowing Economic Man and more reliance on psychology, sociology, polical science, and anthropology.

This suggests that a kind of convergence may be occurring between certain segments of research geography and geographic education. Interest in the spatial aspects of human behavior generally relates closely to the interest of the geographic educator in the behavior of students faced with particular kinds of classroom situations. Interest in the spatial aspects of communication flows and information networks generally is not unlike the interest that the geographic educator has in just what kind of communication or information transfer actually occurs in classrooms. Similarly, interest in the cultural diffusion process as it operates over continents or the globe is directly analogous to interest in the diffusion process as it operates within educational systems and programs. And so on. The experience of the High School Geography Project suggests that we may be moving into a new age for geographic education, an age in which the intellectual problems of geographic education are seen as no less exciting than those of some branches of "substantive" geographic research because they are seen as fundamentally the same.

Of course, some aspects of research trends in geography were not mirrored in developments in the High School Geography Project. There was divergence as well as convergence. The more deductive kinds of geography (example: Christallerian geography) did not fare very well in the project. Central Place hexagons proved to be much more exciting to professional geographers than to high school students. The "systems analysis" approach to geography, with its emphases on feedback, equilibrium states, steady states, paths, and the like may, indeed, be teachable at the 10th grade, but the project did not develop appropriate strategies for doing so. Possibly for very good pedagogical reasons. Indeed, though some of the "mathematical" bent of modern geography comes through in *Geography in an Urban Age*, not much. The current low level of mathematical sophistication of most students of geography at either the high school level or the college level may long defeat a very rapid diffusion of many of the ideas of "mathematical" geography. The solution to this problem lies outside geography itself, in the school system as a whole and the society as a whole.

Though some aspects of contemporary geography proved difficult to put into effective "discovery learning" framework, the overall response of the project to the challenge to develop discovery learning materials was exemplary, both for the curricular movement as a whole and for geography itself.

Summary and conclusions

The challenge of the general curricular reform movement in America was a formidable one for geography. Initially, the High School Geography Project tried to meet the *structural* aspect of the challenge. Though some significant contributions toward articulating geography's structure were made, interest gradually came to focus on the other major aspect of the challenge, *discovery learning*. Here the project response was one of the most outstanding in the whole curricular movement. The highly stimulating course produced, *Geography in an Urban Age*, has a strong behavioral emphasis, paralleling some recent trends in geographic research.

If the High School Geography Project is a representative sample of what can be done through *temporary* organizations to overcome the rigidity and lethargy which seems to characterize our Organizational Age, then we can look forward to the future with much hope. Similar temporary organizations of this general type will no doubt be needed again and again in the future. One of the most distinctive aspects of the project experience is that the professional geographers involved in it were able to grow and to change so that they could eventually to see beyond their own professional needs to the needs of mankind whom we professionals claim to serve.

References

ABLER, R., ADAMS, J. and GOULD, P. (1971) *Spatial Organization*; (Prentice-Hall, Englewood Cliffs, New Jersey).

Association of American Geographers (1965) *Geography in Undergraduate Liberal Education: A Report of the Geography in Liberal Education Project*; (Washington, D.C.), 66 p.

BRUNER, J. S. (1963) *The Process of Education*; (Vintage Books, Random House, New York), 97 p.

COREY, K., et al (1971) *The Local Community: A Handbook for Teachers*; (Macmillan, New York).

DICKINSON, R. E. (1969) *The Makers of Modern Geography*; (Praeger, New York), 305 p.

GARDNER, J. W. (1961) *Excellence: Can We Be Equal and Excellent Too?*; (Harper and Row, New York).

HARVEY, D. (1969) *Explanation in Geography*; (Arnold, London), 521 p.

High School Geography Project (1970) *Teacher Education Kit: Using Simulation to Involve Students*; (Boulder, Colorado).

HILL, A. D. (1970) Strategies of the High School Geography Project for the Colleges: A New Heresy; *Journal of Geography* 49, 544-51.

Joint Committee on Education of the Association of American Geographers and the National Council for Geographic Education (1962) *Advisory Paper for Teachers Associated with the High School Geography Project*.

KOHN, C. F. (Ed.) (1964) *Selected Classroom Experiences: The High School Geography Project*; (National Council for Geographic Education, Normal, Illinois).

KUHN, T. (1962) *The Structure of Scientific Revolutions*; (University of Chicago Press, Chicago), esp. 169-172.

PATTISON, W. D. (1964) The four traditions of geography; *Journal of Geography* 43, 211-16.

PATTISON, W. D. (1970) The Producers: A Social History; *From Geographic Discipline to Inquiring Student, Final Report on the High School Geography Project*; (Washington, D.C.), 57-169.

PATTON, D. J. (Ed.) (1970) *From Geographic Discipline to Inquiring Student, Final Report on the High School Geography Project*; (Washington, D.C.), 102 p.

THOMAS, E.N. (1964) Some comments about a structure of geography with particular reference to geographic facts, spatial distribution, and areal association; *Selected Classroom Experiences: High School Geography Project*. (Later developed more fully by Thomas as a provisional unit for *Geography in an Urban Age*. During the life of the project, a number of such provisional units were produced but not included in a highly visible way in the final course which emerged.)

WHITTLESEY, D. (1954) The regional concept and the regional method; In James, P. E. and Jones, C. F. (Eds.), *American Geography Inventory and Prospect* (The Association of American Geographers, Syracuse University Press), 19-68.

PART VI

Ethical

14 · Ethics and logic in geography

W. BUNGE

No contradiction exists between ethics and logic in geography or anything else. It is illogical to be unethical; the thought of Spencer, among others. Aquinas and others long ago turned it around to "It is unethical to be illogical". What great commonality exists between science and people, between black slum dwellers in Detroit and nuclear physicists in Berkeley? It is the issue of survival. This is the issue that ties all, logic and ethics, operationally.

Geography's role in human survival

There is only one human purpose which is for its own sake; that is the collective survival of the species *Homo sapiens*. It was only by the processes of collective survival that mankind or any other life form came into existence. It is as silly to ask what is the purpose of human life as it is to ask what is the purpose of ant life. The difference between ants and humans is not that they are not tied to the same iron laws of Darwin but that only the humans can understand the iron laws by which they must live. Men cannot repeal the laws of Darwin any more than they can repeal the laws of Newton. Mankind exists for its own sake, to continue itself.

This Law of Life means that nothing else human exists for its own sake. Everything, directly or indirectly, consciously or sub-consciously, exists for the sake of collective existence, for survival. How, then, are very clear subjective attributes of "doing things for their own sake" explained? The trade of farming, of growing food, is clearly a survival-related industry in a most direct way. Yet, do farmers say to themselves, "I'm plowing this field so that the race survives." Such a thought pattern, an obsessive concentration on the ultimate purpose of farming, would reduce the quality of farming, thus reducing survival chances. Instead, the farmer takes pride in his work for its own sake, enters plowing contests, thinks of the money his work will bring him, which improves his concentration on the task at hand. All labour develops a pride in work if it is basically contributing to long-term survival. Is "useless mathematics" really "useless"? The very mathematicians that admonish students to follow "useless" mathematics are filled with stories of how the pursuit of "useless" mathematics had led to the most amazing practical advances. This is a low order paradox, one of sheer semantics. Mathematicians, of all people, should correct their sophomoric error of calling mathematics "useless" and "usefull" in the same breath. In mathematics, one does not pursue "idle curiosity" by "hardworking curiosity". The solution to "puzzles" in mathematics and science requires the highest order of concentration; very hard work.

Since nothing went into the brain but survival for its own sake, does it follow that the brain can think nothing but survival thoughts? No. No part of the human body can afford to be perfect. If the brain were perfect, it would lower survival odds; it would be too big — enormous relative to the stomach or other parts — so in the odds of the construction of life it is more efficient not to pursue perfection. Indeed, perfection would be collective suicide. We therefore suffer from diseased lungs and thoughts.

By what process do we sort out survival and non-survival thoughts? Many trades have developed with this labour, two very important ones being science and art. We hold geography so fragile as if a missed citation will destroy us. Geography is a very clear survival subject. What is the deadliness of being lost individually in the desert? What is the collective deadliness of being lost as a species? For instance, the Soviet Union is not "way over" in Russia, or even "ninety miles off our shores" in Cuba, but directly up. It has always been our job to help mankind find itself.

Geography and political interference

At the end of the article, I will suggest more survival-related work for our trade; hopefully we will enter into dignified, co-operative effort. But the next subject that must be raised is most disagreeable. It has to do with our relationship as craftsman to the rest of society, especially what is commonly thought of as "political" apparatus.

Those jobs that are truth-telling are essentially a conversation between the collective human race and reality, rather than a conversation between men as in conventional jobs. Just what type of personalities are attracted to truth telling is not the point. The point is that out of the nature of the day-to-day effort of being a scientist or an artist certain "non co-operative" or "disruptive" attitudes develop relative to the co-operative effort that most people make. The truth-telling jobs are similar to the jobs on a ship that relate the ship to the ocean itself, such as the radar man, or the navigator, or the man in the crow's nest. These jobs do not require the same abilities as that of the officer who must harmonize all the people on board the ship and carry the ultimate responsibilities. If a lowly sailor in a crow's nest sees an iceberg and the captain orders him to not see it for fear of panicking the ship, what happens to the poor sailor in the crow's nest? And a little later, what happens to the poor ship – including the poor captain?

This does not mean that truth telling is not supposed to be risky. None of the human race is going to be truly safe until all are. Nor does this mean that truth telling is supposed to be absolutely free – like the principle of "academic freedom" seems to imply. None of us will be totally free until all of us are. Nor does this mean that truth telling is supposed to be understood by others. There are relatively few jobs in truth telling and many, many jobs in co-operative organizational work. Why should the few expect the many to "understand" us? But it is important that we not compound the world's understandable, if maddening, misunderstanding of our work, by errors of our own. That is, we have to struggle against the errors that the world makes relative to us, but can maturely expect no more than a compromised understanding and sympathy in spite of our best effort to make our point understood. But inside the trade, among ourselves, we sometimes needlessly drag in the errors of the outside world in order to attack each other. This we can correct and therefore should do immediately, and while I am not hopeful that this article will have much effect on the behavior of the generals in the world, it might do something for our brotherhood of geography.

The "politics" or any other mental state of the truth teller expresses itself in the man's labour. It is not possible to tell if a

plumber is a Communist or Fascist by examining his work. It is
possible to tell that about a poet in his work. Truth tellers wear their
minds on their sleeves by definition of their trade. Politicians know
this and so they are constantly picking over the minds of the truth
tellers work to find "subversive thoughts". I speak from a
highly-developed set of what is normally thought of as "political"
opinions myself. I admire most of the social institutions of the
Soviet Union and wish to adopt what I consider their gains to my
country. I am extremely anti-Capitalist, having been raised in their
secret world. That is, from what I see outside and inside my country,
I have a clear opinion about "political matters". To many of my
colleagues, these last sentences seem like some terrible admission of
political bias in my work. Far from it. Who does not have a
"political point of view", including especially those supercilious
geographers who claim that their political view is to have none? —
that is, who accept the established politics without admitting to the
existence of it. My "confession" is not even an illustration of my
honesty. It is an illustration of the nature of the mental labour called
geography. Geography has been the overwhelming force in leading
me to such a deep "political" position. Having lived and struggled in
this black neighbourhood of Detroit called Fitzgerald, from which I
write, how could I avoid directing my attention to this region? And
in the dialectics of the work, the commerce between the labour and
the worker, how else could my work not help shape what I
think — and therefore, as a geographer, shape me? In the trades of
truth telling, there is no such thing as a "dedicated" Communist or
Capitalist; there is only a convinced one. All truth telling, by its
definition, must be voluntary work. People with political preconceptions, with primary loyalties to the trades of organization, write,
paint, or put on plays that "lack conviction". This is what the
organizers who try and stamp out political deviations do not
understand. Artists and scientists have to be freely won. If they are
not freely won, they are not artists and scientists. Power can force
an appearance of agreement in the trades of art and science, just as
power can force the appearance of love, but only love buys love and
only freedom buys truth.

  There exist two ways to catch butterflies. One way is with a net,
a bottle of chloroform and mounting case. Such a butterfly, with a

pin through its heart, is a trophy, not a living butterfly. The other way to catch a butterfly is to be a beautiful flower and take the chance that the buttefly might fly away if it changes its mind. And if it does fly away, do not reach for the bottle of chloroform but be self-critical, try harder to be a beautiful flower. This is the ideal work relationship between flowers and butterflies and politicians and truth tellers. It is not the full story, and it is a little too delicate for the times. In the future, in gentler times, it should be our goal. But to look at the relationship at the harsh extreme, what can the politicians do if they ruin science?

If I personally had the power to discipline geography, would I do it? Yes. Well, if I would impose myself on the trade, what right do I have to complain of the use of political power against me? In the first place, it must be very clear just what I would ruthlessly purge. I would ruthlessly purge anything that was directly and clearly anti-children, since I consider being anti-children is the same thing as being anti-life, and therefore a biological pathology – a disease. I would protect the minds of children from the minds that hate them. I have written two geography books on this theme and have worked for over five years in frantic effort in this direction, so it can hardly get a fair presentation in these few paragraphs, so I will not attempt it. In America, this certainly means removal of racist materials in geography books. But I would remove with power nothing else other than clearly anti-children material, nor would I "expand" the definition of "anti-children" beyond just that. It would not be some rubber-stamp device to cover anything I, or like-minded people outside geography, might want to impose. The second question has to do with method. Free from the outside political pressures American geographers are so clearly under, the overwhelming majority of American geographers can be trusted to be pro-children. Ultimately, it is the geographers themselves, completely free of outside instruments of the state, that should decide not only what is anti-children in American geography but much more creatively, how to be pro-children in ways yet unimagined. Geographers must be responsible to the mothers and fathers of the children they are harming. The parents, and their instruments, have the right to protect their children's minds where geographers wilfully fail. The third question is "Yes, that might be what you want, but once the workers seize the instruments of the state, what is to assure us that these guidelines will not be broken down by direct and destructive political interference, as did Stalin in the Soviet Union?" This question has to be answered in two ways. I personally obviously cannot guarantee anything, but geographers as a group could do a much better job. That is, if geography takes and steadfastly holds to a deeply-humanist effort, it can "poison the minds" of the politicians not only in providing humanist material but the message as to what is needed, what "work conditions" are needed (free inquiry) for the continued production of such material. Even this does not absolutely assure geographers of free trade conditions. The question then becomes, "What is the alternative? Should we continue heavily in military and spy work?" With the human race facing potential annihilation, I guess even geographers face risks.

When a man is defined out of his work, out of his labor, he in many ways has been defined out of existence itself. To say of a man's work "that is very interesting, but that is not geography", is such an aggression and should be resisted as seriously as saying "that is not permissible geography". This is the error of 'establishment' geographers, not that they in turn can be defined out of existence. To avoid contradiction, it is vital not to dismiss such workers as somehow "illegitimate" geographers, especially if they dismiss others. Nor is this some "noble" generosity; it is done to arouse trade sentiment against this sort of "reasoning". It is essential that the central direction of geography comes from more working within the trade. The work of geography is the gyroscope of its work, not someone else's work imposed from without. Our work, geography, is everything. It is what we have in common, what we do and what we are. In that we do not labor, are lazy, expedient, opportunistic, fearful; in that degree, we are not geographers. Much of the political manhandling we receive we ask for. How many geographers string out their day around endless coffee breaks and complain about the nature of geography rather than the nature of geographers? The internal ethics and logic of geography is to do our labor, self-consciously resisting injury to children.

"Mathematical" versus human geography

The substance, as opposed to the politics, of political attack shifts the ground back into our trade, geography, and need not be answered in such strident terms. Al'brut, along with many American geographers, does not believe in the humanism of theoretical or so-called "mathematical" geography. He writes:

These and similar ideas are incorporated by Bunge into his mathematical geography, which, without foundation, he treats as theoretical geography. The equating of terms is not accidental since all theory, in his view, is reduced to mathematization demonstrating the unity of geography and the harm of separating it into economic and physical geography.
(Al'brut, M.A., Book Review, *Soviet Geography* 1970, 141)

Similar objections were raised earlier in America. In an unsigned article in *New University Thought*, the following passage yields the flavor:

Now the consensus of opinion (sic) is that geographers have only recently begun to unmask principles of 'spatial interrelations'. We are, it is maintained, on the threshold of understanding the nature and order of things in space.
The question I raise is whether or not these findings permit an investigation and evaluation of the human condition and ultimately the attainment of the social good.
(Unsigned, *New University Thought* 3, No. 1, 1963, 64)

I replied to this article in the next issue ("A Geography for Humans", *New University Thought* 1963, 62-4) with a hypothetical illustration of the use of theoretical geography in locating children's "tot lots" around the city of Detroit to better serve the children. An extraordinary opportunity in this respect occurred when the City of Detroit re-districted the schools in 1969 to yield better "community control", but this opportunity was let slip. However, geographers in the Detroit Geographical Expedition and Institute, working directly for a large organisation of black industrial workers, prepared a plan with the most sophisticated geographical techniques. The computer at Queen's University, Kingston, Ontario, under the guidance of the English mathematical geographer, John Shepard, and the mathematical programmer, Michael Jenkins, was used to investigate every possible combination of legal districts, evolving a sophisticated algorithm in the process to give the most sympathetic authority to the school children. This study led to a compromise with the School Board which would have been impossible without computer-based research.

Other "mathematical" geographers in the United States have been reacting against the irrationality of centralized planning with true rationality. The most irrational feature of any American city is the fact that children are hungry in the midst of abundant food right in their neighborhoods. Planning for the "power structure of the outs" is therefore the most rational planning since it is in the interests of the slum dwellers to remove slums, the most irrational urban feature. All over the continent, geographers, including "mathematical" geographers, are bending their trade toward an alliance with the under-privileged. In Philadelphia, two "mathematical" geographers, Julian Wolpert and Thomas Reiner have been instrumental in resisting an expressway, and have compiled a case file of misplanning across the continent. At Harvard, William Warntz's Center for Computer Graphics has been directly concerned with combating inhuman planning. In Detroit a program of computer mapping is being prepared which will be of great use in the rapid mapping of election situation, unemployment data, and indeed of anything requiring quick analysis. The "mathematical" geographer, Gerald Karaska at Clark University, has consistently supported the Detroit "Opens Admission" program. Richard Morrill, the "mathematical" geographer at the University of Washington in Seattle, Allen Pred at Berkeley and many others have been in the forefront against racism in American geography. Walther Isard, the founding father of Regional Science, is an ardent worker for peace. Andrew Karlin, Herman Porter, Richard Guyot, John Borchert, and on and on, are "mathematical" geographers in the struggle for humanism. Brother Al'brut, you do not know whereof you write!

This does not mean that some "mathematical" geographers have not used the work to obfuscate, to hide behind a facade of logic in pursuit of personal careers, or to divert legitimacy from "value judgement" and "prejudiced" socially-applied geography to their version of "pure" science. Sometimes even well-meant efforts by mathematical geographers fail to achieve their objectives; as did the well-known *Atlas of Economic Development* by Norton Ginsburg and Brian J. L. Berry, due to a technical error. The atlas is fine, humane, in conception, but the scaling of all the map categories is a confusion of technique. It is not only impossible to use the maps with "plain ordinary people", but it is impossible for "professional" geographers to easily understand them. It is this individual geographer's opinion that in this case the "mathematical" geography obscured the work rather than clarifying it. So the humanist geographers, be they Al'brut or anyone, do not lose their argument hands down but present us with a half-truth, which is always difficult to answer and requires balance in reply.

Geography as a unifying force between Man and Nature

Having spent much of the labor of this article trying to correct work attitudes, the remainder will break new ground. In both the United States and the Soviet Union, there has existed resistance to a unified theory of geography. Anuchin produced a book entitled *Theoretical Geography* which called for unity between "economic" (our "human") and "physical" geography, and elsewhere wrote:

. . . The development of society on earth was regarded as isolated from the development of nature, and *vice versa*. And yet the life of society proceeds in close conjunction with the development of the geographic environment. The environment is not only and simply external nature. It is a distinct part of external nature, a distinct form of matter that constitutes a material unity with society. The influence of the geographic environment is not only the influence of earth's nature, but also the influence of the human labor of all the 'societies' that have existed on earth and whose work has been embodied in nature.
(Anuchin, V.A., A Sad Tale About Geography, September 1965, *Soviet Geography* VI, No. 7, 29)

My version of Theoretical Geography implies such a unity, since the basic language of the subject, similar to abstracted languages in other disciplines, has to do with Schaefer's "spatial relations" not the nonspatial properties of the objects. For instance, a dendrite is simply a centrally-placed line similar to Christaller's centrally-placed points. Of what could the underlying pattern be a map?

Is this necessarily a river? How about a man-made sewer system? Perhaps if your local sewer system does not have this pattern, it should have for maximum efficiency. In a book on which William Warntz and I have been working, seemingly for ever (ten years) but nearing completion at last (*Geography: An Innocent Science*; John Wiley), I wrote most of the first draft in traditional physical-human form, indeed, following the traditional "topical" (misnamed, as Warntz points out, "systematic") chapter headings ("climate," "landforms", etc). The pattern of the book proved so repetitious that it forced a re-drafting to achieve a systematic unity. (So-called "systematic" texts, Warntz points out, have "no system" in that you could just as easily read the last chapter first, so they are "traditionally arranged topics". In the forthcoming Warntz-Bunge book, it is indeed systematic and you must read the chapters in order, like a physics book).

But if efficiency has forced a unity on geography on "abstract" grounds, there are some *extremely* human grounds for the same unity. As Hartshorne wrote:

> The traditional organization of geography by topics into two halves, "physical" and "human", and the division of each half into sectors based on similarity of the dominant phenomena in each, is of relatively recent origin and has proven detrimental to the purpose of geography.
>
> (Hartshorne, *Perspective on the Nature of Geography* 1959, 79)

The definition of geography as a study of the earth's surface as the home of man is probably the most widely accepted in America. I prefer it as a general definition of the trade to my own "Geography is the science of location" since my own is not inclusive enough to cover all our work. I also do not believe in single definitions since I feel a labor can be described from more than one point of view, many-sidedly. Differences are not necessarily antagonistic; they are often complementary. Aside from its virtue of wide acceptance, the phrase "the earth's surface as the home of man" raises the question, "Is, in fact, the earth's surface man's home or his graveyard?" This, in turn, raises questions about man-made regions. As a whole, the earth already is a man-made region, profoundly affected by radioactivity for instance. This, in turn, raises questions about a "natural region". We must move forward to nature and build natural man-made regions, "natural" in the sense that they are regions of long-term survival homes for the species *Homo sapiens*. It is not ecology but geography that the world needs, since ecology is "physical environment minus mankind". It is geography that puts man in nature. The American philosopher who philosophizes over geography, Vernon Dolphin, puts it this way:

Many of us are taught but one side of geography. We are taught to locate foodstuffs, capitals and weather patterns on the map. We get the impression that locating things is the main concern of the geographer. It is one concern. He is also concerned to describe how man can go from one place to another to do and see interesting things; how some men help others by growing needed things; how a culture helps man survive and thrive in a given terrain; how men share or do not share a common set of conditions such as weather, work, or goads to war. More than an inventory of man and things, geography at heart is a concern to depict man's relation to man upon earth, as though earth were his home.

(Vernon Dolphin, *The Earth is the Home of Man*; Education Department, Harvard University, 1970, 4)

The full geographic arguments urging that we build natural regions cannot be placed in such a short article as this but will be found in the last part of an article called *The Geography of Human Survival* (In preparation).

The dichotomy between "physical" and "human" ("economic") geography forces the question to become "man against nature" rather than "man in nature". Frederick Engels wrote the following:

> In short, the animal merely uses external nature, and brings about changes in it simply by his presence; man by his changes makes it serve his ends, masters it. This is the final, essential distinction between man and other animals, and once again it is labor that brings about this distinction.
>
> Let us not, however, flatter ourselves overmuch on account of our human victories over nature. For each such victory it takes its revenge on us. Each of them, it is true, has in the first place the consequences on which we counted, but in the second and third places it has quite different, unforeseen effects which only too often cancel the first. The people who, in Mesopotamia, Greece, Asia Minor, and elsewhere, utterly destroyed the forests to obtain cultivable land never dreamed that they were laying the basis for the present devastated condition of these countries, but removing along with the forests the collecting centers and reservoirs of moisture. When the Italians of the Alps used up the pine forests on the southern slopes, so carefully cherished on the northern slopes, they had no inkling that by doing so they were cutting at the roots of the dairy industry in their region; they had still less inkling that they were thereby depriving their mountain springs of water for the greater part of the year, and making it possible for them to pour still more furious torrents of it on the plains during the rainy seasons. Those who spread the potato in Europe were not aware that with these farinaceous tubers they were at the same time spreading scrofula. Thus at every step we are reminded that we by no means rule over nature like a conqueror over a foreign people, like someone standing outside nature but that we, with flesh, blood and brain, belong to nature, and exist in its midst, and that all our mastery of it consists in the fact that we have the advantage over all other creatures of being able to know and correctly apply its laws.
>
> (Frederick Engels, *The Part Played by Labour in the Transition from Ape to Man*, 1876, pp. 18-19)

I agree with Engels. Mankind starts out trying to conquer nature and ends up learning to live with it, or else. Physical and human geography had better not be split.

Wilbur Zelinski, the American geographer, has suggested a set of "voluntary regions" which seem most reasonable to me. Let us start working man into nature by building regions, homes of mankind! Americans as a biological segment of the species have been through so much this last decade that there is danger that the American young will become suicidal — murderous in mass. The whole species could collectively just shrivel up and die of radioactivity if massive hope is not injected immediately. So let us start to build Paradise, not as escapists, but tied into today's conditions. For Detroit, perhaps this might be an island in the Caribbean to which the deprived children, the poorest orphans, of Detroit could be brought. As we move from the dictatorship of the proletariate, to the dictatorship of the children; from building Paradise to Paradise itself; from Socialism to Communism, from unnatural to natural man; it would be helpful even under the oppression in America to obtain as clear a dream of our goal as it is materially possible to substantiate.

What role does theoretical geography play in all this humanism? Well, look at the operational embarrassment of our pre-theoretical stage. We are "the integrating science", so we call upon our co-workers in geology, sociology and so forth to discuss planning for a region, or even the lesser labor of just understanding a region with no ambitions humanly to improve it. The room fills with people who sit nicely around a table and begin to go to work. In the pre-theoretical days, when our science was still merely classifactory (regional) and particularizing (place-naming), we could never enter into the discussion of analysis or prediction. We had nothing to say. We could provide maps, and perhaps a general "feel of the region" and we had the imagination to call the meeting, but then a silent embarrassment overcame us.

Today, if geographers were to assemble a team of explorers to build natural regions, the "integrating science" has something to say continuously, long after presenting regional and cartographic locational material. Every time something has to be located, it will be our role to do the labor of proper location and we will sit in the full dignity of our labor among predictive co-workers from other fields of knowledge.

Since science must be done, not simply thought about, I would like to make the following suggestion: that a group of geographers from non-Socialist countries and a group of geographers from Socialist ones hold a joint conference to consider building a set of sub-regions for man's home, which eventually will become the kind of regions that all mankind will occupy. It might be most scientifically efficient to try two very different regions of the earth, say the Western effort in the Caribbean and the Eastern in Zelinski's "montane" voluntary region.

Why in the countryside? Why not in the cities where people already exist? Is this not escapism? First, we already are working in cities. The bulk of the effort of human exploration is directed by definition to where the bulk of the people live and in North America that is the cities. We must struggle to make the regions that are the cities more natural, that is, more collectively survival prone. But improving slums is not enough. It does not set high enough standards. Mankind has always been a shifty species. He poisons his nest. His presence becomes deadly to himself so that he is continually constructing and reconstructing but he needs a "region forever". He needs to establish places where the long-time survival chances of the species are good. Children are the weakest necessary connection with life. Therefore if children are prospering in a space, so is the species. Contrarily, if the children are dying, as in American slums with high infant mortality rates, or in Vietnam, then these are regions in which the species itself is dying. If such regions become large enough, say by the spread of radioactivity, to cover the whole planet then the species itself is doomed. You do not have to create an environment any worse than being deadly to children to destroy the species. While slums and aggressive wars must be struggled against day-to-day, we must also not abandon the strength that comes from not just a vision of the necessary world, but of a much better one. To put it less grandly, why are all the best spots in the world in the hands of the jet set? Surely we should attempt to build prototype paradises as part of the struggle against slums. For instance, should not the children of the slums be introduced, even for short vacations, to better regions? Paradise should be made the property of all, especially for the poor.

Does all this sound utopian? Good. Science is often accused of being impractical. Some "new ideas" sound new for as long as a decade. In the countryside we can project the new cities. It is not suggested that the geographers stay permanently in the new regions. They must work most of the time where the people are, back in the cities. But geography has very little resources so we must construct regions where we can and for now that means the countryside.

May the earth be filled with happy regions.

Printed in the United States
by Baker & Taylor Publisher Services

Printed in the United States
by Baker & Taylor Publisher Services